Mathematics for Industry

Volume 35

Aims & Scope

The meaning of "Mathematics for Industry" (sometimes abbreviated as MI or MfI) is different from that of "Mathematics in Industry" (or of "Industrial Mathematics"). The latter is restrictive: it tends to be identified with the actual mathematics that specifically arises in the daily management and operation of manufacturing. The former, however, denotes a new research field in mathematics that may serve as a foundation for creating future technologies. This concept was born from the integration and reorganization of pure and applied mathematics in the present day into a fluid and versatile form capable of stimulating awareness of the importance of mathematics in industry, as well as responding to the needs of industrial technologies. The history of this integration and reorganization indicates that this basic idea will someday find increasing utility. Mathematics can be a key technology in modern society.

The series aims to promote this trend by 1) providing comprehensive content on applications of mathematics, especially to industry technologies via various types of scientific research, 2) introducing basic, useful, necessary and crucial knowledge for several applications through concrete subjects, and 3) introducing new research results and developments for applications of mathematics in the real world. These points may provide the basis for opening a new mathematics-oriented technological world and even new research fields of mathematics.

To submit a proposal or request further information, please use the PDF Proposal Form or contact directly: Swati Meherishi, Executive Editor (swati.meherishi@springer.com).

More information about this series at https://link.springer.com/bookseries/13254

Jin Cheng · Xu Dinghua · Osamu Saeki ·
Tomoyuki Shirai
Editors

Proceedings of the Forum "Math-for-Industry" 2018

Big Data Analysis, AI, Fintech, Math in Finances and Economics

Springer

Editors
Jin Cheng
School of Mathematical Sciences
Fudan University
Shanghai, China

Xu Dinghua
Shanghai University of Finance and Economics
Shanghai, China

Osamu Saeki
Institute of Mathematics for Industry
Kyushu University
Fukuoka, Japan

Tomoyuki Shirai
Institute of Mathematics for Industry
Kyushu University
Fukuoka, Japan

ISSN 2198-350X ISSN 2198-3518 (electronic)
Mathematics for Industry
ISBN 978-981-16-5575-3 ISBN 978-981-16-5576-0 (eBook)
https://doi.org/10.1007/978-981-16-5576-0

This Springer imprint is published by the registered company Springer Nature Singapore Pte Ltd.
The registered company address is: 152 Beach Road, #21-01/04 Gateway East, Singapore 189721, Singapore

Organization

Forum "Math-for-Industry" 2018
- Big Data Analysis, AI, Fintech, Math in Finances and Economics -
November 17–21, 2018

Organized by

Fudan University, Shanghai, China

Scientific Committee

Tatsien Li, Academician of CAS, Fudan University, China
Shige Peng, Academician of CAS, Shandong University, China
Masato Wakayama, Executive Vice President and Senior Vice President of Kyushu University, Japan
Bob Anderssen, CSIRO, Australia
Huaxiong Huang, York University, Fields Institute, Canada
Yasuhide Fukumoto, Institute of Mathematics for Industry, Kyushu University, Japan
Tomoyuki Shirai, Institute of Mathematics for Industry, Kyushu University, Japan
Naoyuki Ishimura, Chuo University, Japan
Faouzi Triki, University of Grenoble, France
Victor Isakov, Wichta University, USA
Masahiro Yamamoto, The University of Tokyo, Japan
Gang Bao, Zhejiang University, China
Jin Cheng, Fudan University, China
Jijun Liu, Southeast University, China

Organization Committee

Tatsien Li (Chair), Academician of CAS, Fudan University, China
Masato Wakayama, Executive Vice President and Senior Vice President of Kyushu University, Japan
Bob Anderssen, CSIRO, Australia
Jin Cheng, Fudan University, China
Yasuhide Fukumoto, Institute of Mathematics for Industry, Kyushu University, Japan
Dinghua Xu, Shanghai University of Finance and Economics and Zhejiang Sci-Tech University, China
Shuai Lu, Fudan University, China
Wenbin Chen, Fudan University, China

Keynote Speech

Shige Peng, Academician of CAS, Shandong University, China

Plenary Talk

Samuel Drapeau, Shanghai Jiaotong University, China
Weiguo Gao, Fudan University, China
Keiichi Goshima, Institute for Monetary and Economic Studies, Bank of Japan, Japan
Lê Minh Hà, Vietnam Institute for Advanced Study in Mathematics, Vietnam
Naoyuki Ishimura, Chuo University, Japan
Tadashige Iwao, Fujitsu Limited, Japan
Sergey Kabanikhin, Institute of Computational Mathematics and Mathematical Geophysics of the Siberian Branch of the RAS, Russia
Takayuki Osogami, IBM Research—Tokyo, Japan
Pan Qin, Dalian University of Technology, China
Jun Sekine, Osaka University, Japan
Taiji Suzuki, The University of Tokyo, Japan
Shigeo Takahashi, University of Aizu, Japan
Eric Ulm, Victoria University of Wellington, New Zealand
Jonathan Wylie, City University of Hong Kong, HKSAR, China
Dinghua Xu, Shanghai University of Finance and Economics and Zhejiang Sci-Tech University, China
Songping Zhu, University of Wollongong, Australia

Young Researchers Talk

Nathan Gold, York University, Fields Institute, Canada
Ling Guo, Shanghai Normal University, China
Guanghui Hu, Beijing Computational Science Research Center, China
Maxim Shishlenin, Sobolev Institute of Mathematics of the Siberian Branch of the RAS, Russia
Yutaro Kabata, Institute of Mathematics for Industry, Kyushu University, Japan
Xiliang Lv, Wuhan University, China
Taku Moriyama, Institute of Mathematics for Industry, Kyushu University, Japan
Min Zhong, Southeast University, China

Program at a Glance

<table>
<tr><th></th><th>Nov. 16
(FRI)</th><th>Nov. 17
(SAT)</th><th>Nov. 18
(SUN)</th><th>Nov. 19
(MON)</th><th>Nov. 20
(TUE)</th><th>Nov. 21
(WED)</th></tr>
<tr><td>09:30 – 09:45</td><td rowspan="7"></td><td>Opening Ceremony</td><td rowspan="3">Naoyuki Ishimura
Jun Sekine</td><td rowspan="3">Songping Zhu
Keiichi Goshima</td><td rowspan="3">Shigeo Takahashi
Dinghua Xu</td><td rowspan="3">Taiji Suzuki
Tadashige Iwao</td></tr>
<tr><td>09:45 – 10:45</td><td>Shige Peng</td></tr>
<tr><td>10:45 – 11:00</td><td>Group Photo</td></tr>
<tr><td>11:00 – 11:30</td><td colspan="5">Coffee Break</td></tr>
<tr><td>11:30 – 12:15</td><td rowspan="2">Sergey Kabanikhin
Ling Guo</td><td rowspan="2">Samuel Drapeau
Nathan Gold</td><td rowspan="2">Eric Ulm
Taku Moriyama</td><td rowspan="2">Pan Qin
Yutaro Kabata</td><td>Takayuki Osogami</td></tr>
<tr><td>12:15 – 12:45</td><td>Closing Ceremony</td></tr>
<tr><td>12:45 – 14:00</td><td colspan="5">Lunch</td></tr>
<tr><td>14:30 – 15:15</td><td rowspan="7">Registration</td><td rowspan="2">APCMfI Executive Meeting</td><td rowspan="3">Jonathan Wylie
Weiguo Gao</td><td rowspan="2">Lê Minh Hà
Maxim Shishlenin</td><td rowspan="7">Poster Session</td><td rowspan="8"></td></tr>
<tr><td>15:15 – 15:45</td></tr>
<tr><td>15:45 – 16:00</td><td>Free Discussion</td><td rowspan="2">Coffee Break</td></tr>
<tr><td>16:00 – 16:15</td><td rowspan="3">APCMfI Meeting</td><td rowspan="2">Free Discussion</td></tr>
<tr><td>16:15 – 16:45</td><td rowspan="3">Guanghui Hu
Min Zhong
Xiliang Lv</td></tr>
<tr><td>16:45 – 17:15</td><td rowspan="2">IMI Meeting</td></tr>
<tr><td>17:15 – 17:45</td><td>Free Discussion</td></tr>
<tr><td>17:45 – 19:30</td><td>Dinner</td><td>Banquet</td><td colspan="3">Dinner</td></tr>
</table>

Preface

We are in the era of big data, and every day, or even every second, a huge amount of data is being accumulated or collected in our whole world. Such collection of big data has become possible because of the development of computational technology and Internet. Now our most important problem is how to get useful information from such a complex set of data. It is natural to see that statistics plays an essential role in analyzing such big data. Furthermore, mathematics can provide more and more techniques and technologies for such purposes.

Another key word that leads today's world is "AI," artificial intelligence. This is naturally based on the machine learning, especially the deep neural network (DNN) techniques, and is now spreading in vast areas in our daily life. On the contrary, it has also been pointed out that the theoretical reason why such AI systems work so efficiently is still not clear. It is believed that mathematics will overcome such a problem.

These two key terminologies, big data analysis and AI, are now indispensable in finances and economics in the world. Thus, without mathematics or statistics, we cannot talk about today's society.

In these circumstances, we organized the conference, Forum "Math-for-Industry" 2018 (FMfI2018), at Fudan University, Shanghai, China, during November 17–21, 2018, for which the unifying theme was "Big Data Analysis, AI, Fintech, Math in Finances and Economics." We are sure that the theme was very timely and made a big success not only in the industrial mathematics community, but also in industry. This book is the proceedings of the conference and collects together selected papers presented there. The topics covered in the conference are spatial financial risks, foreign exchange markets, option pricing, evolution of copulas, inverse problems connected to financial mathematics, DNN with uncertainty quantification, reinforcement learning, estimation error analysis of deep learning, integration of AI to agriculture, functional clothing design, application of singularity theory, history of modern mathematics in Vietnam, etc.

The contents of this volume also report on productive and successful interactions between industry and mathematicians, as well as on the cross-fertilization and collaboration that occurred. The book contains excellent examples of the roles of mathematics in our society and, thereby, the importance and relevance of the concept Mathematics_FOR_Industry.

We would like to thank the participants of the Forum and the members of the Scientific and Organizing Committees, especially Jin Cheng, Tatsien Li, Shuai Lu, Wenbin Chen and Yu Chen of Fudan University and Dinghua Xu of Shanghai University of Finance and Economics. Without their cooperation and support, we would never have experienced the great excitement and success of the Forum. Moreover, we would like to express our sincere gratitude for the great help of the conference secretaries, Wei Chen, Seiko Sasaguri and Tsubura Imabayashi, during the preparation and organization of the Forum, and also for the proceedings.

On behalf of the Editorial Board of the Proceedings of the Forum "Math-for-Industry" 2018

Fukuoka, Japan
November 2021

Osamu Saeki

Contents

About the Editors

Jin Cheng is a professor at Fudan University, China. He pursued his Ph.D. from the same university. He is currently President of Shanghai Society of Industrial and Applied Mathematics and Director of Shanghai Key Laboratory of Contemporary Applied Mathematics in Fudan University. His research is funded by National Science Foundation of China, the Ministry of Science and Technology of China and Shanghai Municipal Government. Dr Cheng is also on the editorial board of several scientific journals. In 1999, he received the ISACC Young Scientist Award from the International Society for Analysis, its Applications and Computation (ISAAC) in Berlin, Germany. His current research interests concern the inverse problems for partial differential equations, mathematical modeling and analysis, regularization methods for ill-posed problems.

Xu Dinghua is a Professor at Zhejiang Sci-Tech University (ZSTU) and Shanghai University of Finance and Economics (SHUFE). He pursued his Ph.D. degree in Computational Mathematics from Shanghai University, China. He has been awarded the National Distinguished Teacher of China in 2004 and the National Excellent Teaching Achievement Prize in 2014 from the Ministry of Education, China. He was instrumental in the establishment of the School of Mathematics and Informational Sciences (SMIS), ECIT, launched in 2003. His current research interest covers computable modeling, inverse problems for parabolic equations, data modelling for textile material design and for multiscale modeling in catalyst preparation process, numerical computation for partial differential equations.

Osamu Saeki is a distinguished professor at Kyushu University. He received his Ph.D. in Mathematics from the University of Tokyo. He was awarded the Takebe Katahiro Prize 1996 and the Geometry Prize 2015 from the Mathematical Society of Japan. He was involved in establishing the Institute of Mathematics for Industry (IMI), Kyushu University, launched in 2011. He is also engaged in the education of industrial mathematics and is the coordinator of the WISE program "Graduate Program of Mathematics for Innovation", supported by MEXT, Japan. His current research interests concern topology, singularity theory, topology of low-dimensional manifolds, knot theory and visualization of large scale data.

Tomoyuki Shirai received his Ph.D. in Mathematical Sciences at the University of Tokyo in 1996 by a thesis on spectral analysis on discrete Schrödinger operators. He joined the Faculty of Mathematics, Kyushu University in 2004 as an associate professor and became a full professor in 2009. He has been a full professor at the Institute of Mathematics for Industry (IMI), Kyushu University, which was launched in 2011. His research interests are in probability theory, stochastic processes and their applications including the probabilistic aspect of topological data analysis. He is the principal investigator of two Grants-in-Aid for Scientific Research by JSPS on probability theory and related fields and a co-investigator of the CREST project "Topological data analysis for new descriptors on soft matters".

Copula-Based Estimation of Value at Risk for the Portfolio Problem

Andres Mauricio Molina Barreto and Naoyuki Ishimura

1 Introduction

This paper surveys our recent research on the estimation of Value at Risk (VaR) for the portfolio problem.

VaR is one of widely used measures of risk in the field of finance. VaR is the loss in market value over the time horizon T that is exceeded with probability β. Because of its usefulness and clearness, VaR provides a benchmark factor of the risk and plays a principal role in the risk management. Classical methods such as variance–covariance are preferably used so far; however, there is enough empirical evidence which shows that financial returns behave as non-normal distributed random variables with heavy tails and asymmetry, where VaR is apt to be employed. We refer to Duffie and Pan (1997) and McNeil et al. (2005), for instance, and the references cited therein.

Here, we estimate VaR in two ways. One is rather standard and is given as a comparison and/or reference, whose algorithm is based on well-known ARMA-GARCH models, combined with Gaussian mixture innovations (see Lee and Lee 2011). This model may perform better than classical approaches such as variance–covariance and exponentially weighted moving average (EWMA) methods. Using ARMA-GARCH models can capture effects of high volatility on the returns of portfolio. Implementing Gaussian mixture innovations can lead to a more accurate VaR forecasting due to the existence of heavy-tailed and skewed distribution. But it seems that the method fails to capture the relation between variables when considering portfolio of several assets.

A. M. Molina Barreto
Graduate School of Commerce, Chuo University, Tokyo 192-0393, Japan
e-mail: ammolinaba@unal.edu.co

N. Ishimura (✉)
Faculty of Commerce, Chuo University, Tokyo 192-0393, Japan
e-mail: naoyuki@tamacc.chuo-u.ac.jp

J. Cheng et al. (eds.), *Proceedings of the Forum "Math-for-Industry" 2018*,
Mathematics for Industry 35, https://doi.org/10.1007/978-981-16-5576-0_1

The other is copula-based method which is the main purpose of this article. Copulas are well-recognized functions which provide a flexible tool for analyzing the dependence relation among random variables. Because of its readiness for applications, copulas are now customarily employed in various settings. They allow the construction of multivariate distributions even with different margins and dependence structure. See for example Genest and Favre (2007) and McNeil et al. (2005). It is very natural and desirable that the assets are connected with copulas for VaR estimation. Indeed several attempts have been already undertaken, and much progress has been made. See for instance (Fantazzini 2008; Krzemienowski and Szymczyk 2016; Prékopa 2012).

In the present article, we consider a portfolio which is composed of two indexes, namely NASDAQ and Nikkei 225, with the same weight and estimate VaR numerically by the use of both approaches, respectively. Lastly, backtesting shows that copula-based method works better.

The paper is organized as follows: Sect. 2 gives basic definition and properties of Value at Risk and copulas. The determination formula of copula-based VaR is presented in Sect. 3. Empirical study is implemented in Sect. 4. Section 5 concludes with discussion.

2 Preliminary

We briefly make a review on the concept of VaR and copulas for completeness of our presentation. Hereafter, we confine ourselves to the bivariate case for simplicity.

2.1 Value at Risk

Let X, Y be random variables and the portfolio return $Z = \lambda X + (1 - \lambda)Y$ $(0 < \lambda < 1)$. We then see that VaR for Z with the confidence level β $(0 < \beta < 1)$ is given by

$$\mathrm{VaR}_\beta(Z) := F_Z^{(-1)}(\beta) = \inf\{t \mid F_Z(t) \geq \beta\}, \tag{1}$$

where $F_Z(t)$ denotes the distribution function of Z; namely, $F_Z(t) = P(Z \leq t)$.

As to empirical studies, for each observed stock price $\{S_t\}_{t=1}^T$, the daily geometric return r_t is expressed as

$$r_t = \log\left(\frac{S_t}{S_{t-1}}\right) \qquad (t = 1, 2, \ldots, T).$$

VaR represents the maximum expected loss that will not be exceeded with a specified probability β over a predetermined time horizon T.

2.2 Copula

Our additional aspect is concerned with the relationship between our two returns X and Y. It is typical that these variables are assumed to be independent, which makes the situation simpler. However, it may happen that X and Y are nonlinearly related; this is the point we take into account, and we suppose that the nonlinear relation is represented through a copula function. For further details, we refer for instance to (Durante and Sempi 2016; Joe 1997; Nelsen 2006).

Definition 1 *A continuous function C defined on $\mathbb{I}^2 := [0, 1] \times [0, 1]$ and valued in $\mathbb{I} := [0, 1]$ is said to be a copula if the following conditions are satisfied.*
(i) For every $(u, v) \in \mathbb{I}^2$,

$$C(u, 0) = C(0, v) = 0,$$
$$C(u, 1) = u \quad \text{and} \quad C(1, v) = v.$$

*(ii) (*the 2-increasing condition*) For every $(u_i, v_i) \in \mathbb{I}^2$ $(i = 1, 2)$ with $u_1 \leq u_2$ and $v_1 \leq v_2$,*

$$C(u_1, v_1) - C(u_1, v_2) - C(u_2, v_1) + C(u_2, v_2) \geq 0.$$

Next we recall the well-known theorem due to Sklar (1973) in bivariate case. This could be the most important result of the copula theory. It states that any group of univariate distribution can be linked with any copula and a valid multivariate distribution can be defined.

Theorem 1 (Sklar's theorem) *Let H be a bivariate joint distribution function with margins functions F_X and F_Y; that is,*

$$\lim_{x \to \infty} H(x, y) = F_Y(y), \qquad \lim_{y \to \infty} H(x, y) = F_X(x).$$

Then, there exists a copula, which is uniquely determined on Ran $F_X \times$ Ran F_Y, such that

$$H(x, y) = C(F_X(x), F_Y(y)). \tag{2}$$

Conversely, if C is a copula and F_X and F_Y are distribution functions, then the function H defined by (2) is a bivariate joint distribution function with margins F_X and F_Y.

In our empirical study at §4, Student-t, Plackett, and symmetrized Joe-Clayton copulas are used.

3 Determination Formula

Our main analytical observation in this article is about a copula-based VaR. Let X, Y be nonnegative random variables, whose joint distribution function H is represented by a copula C with

$$H(x, y) = P(X \leq x, Y \leq y) = C(F_X(x), F_Y(y)),$$

where $F_X(x) = P(X \leq x)$, $F_Y(y) = P(Y \leq y)$ are marginal distribution functions of X, Y, respectively. We further assume that C has the density c, namely $C(u, v) = \int_0^u \mathrm{d}s \int_0^v c(s, t)\mathrm{d}t$, as well as F_X, F_Y have densities f_X, f_Y, respectively.

We consider a portfolio $Z = \lambda X + (1 - \lambda)Y$ $(0 < \lambda < 1)$ at the confidence level β $(0 < \beta < 1)$ and want to evaluate $\mathrm{VaR}_\beta(Z)$. The determination formula is then stated as follows.

Theorem 2 $\mathrm{VaR}_\beta(Z)$ *can be attained as the unique solution* z^* *for the integral equation*

$$\beta = \int_0^{\frac{z^*}{1-\lambda}} \left(\int_0^{\frac{z^*}{\lambda} - \frac{1-\lambda}{\lambda} y} c(F_X(x), F_Y(y)) f_X(x) dx \right) f_Y(y) dy,$$

so that we see that

$$\beta = \int_0^{\frac{\mathrm{VaR}_\beta(Z)}{1-\lambda}} \left(\int_0^{\frac{\mathrm{VaR}_\beta(Z)}{\lambda} - \frac{1-\lambda}{\lambda} y} c(F_X(x), F_Y(y)) f_X(x) dx \right) f_Y(y) dy. \tag{3}$$

This formula itself is easy to understand; indeed, for simplicity, under the assumption that $F_Z(z)$ is continuous and strictly monotone, we observe that $\mathrm{VaR}_\beta(Z)$ is given by the solution z to the equation

$$\beta = P(Z \leq z) = P(\lambda X + (1 - \lambda)Y \leq z),$$

where the right hand side is reduced to

$$\begin{aligned}
&P(\lambda X + (1 - \lambda)Y \leq z) \\
&= \int_0^z \mathrm{d}s \int_0^s c\left(F_X\left(\frac{t}{\lambda}\right), F_Y\left(\frac{s-t}{1-\lambda}\right)\right) \frac{1}{\lambda} f_X\left(\frac{t}{\lambda}\right) \frac{1}{1-\lambda} f_Y\left(\frac{s-t}{1-\lambda}\right) \mathrm{d}t \\
&= \int_0^{\frac{z}{1-\lambda}} \left(\int_0^{\frac{z}{\lambda} - \frac{1-\lambda}{\lambda} y} c(F_X(x), F_Y(y)) f_X(x) dx \right) f_Y(y) \mathrm{d}y.
\end{aligned}$$

It is to be noted that the formula is already employed in numerical research (see Fantazzini 2008). See also Molina Barreto et al. (2019), where the theorem contains mistakes and which we have corrected here.

4 Empirical Study

We now turn our attention to empirical analysis of estimating VaR.

The database used for our empirical analysis consists of daily geometric return obtained from closing prices for the NASDAQ and Nikkei 225 from August 22, 2013, to August 21, 2018, with a total of 1188 trading days. The data is taken from Yahoo Finance. Table 1 contains descriptive statistics, and Fig. 1 presents plots of both series. The implementation is performed with MATLAB.

Both series present asymmetry and have large kurtosis. In both cases, we can observe the negative value of asymmetry for both series, indicating the likeliness of negative returns, and excess of kurtosis shows fatter tails than the normal distribution. We can also observe the effects of volatility clustering. It would be a good idea to consider model which is different to normal or t-distributed innovations for each series.

4.1 Margins Modelling

To specify a model for each series, we consider ARMA(p, q)-GARCH(r, s) model for asset returns r_t $(t = 1, 2, \ldots, T)$ which is given by (see Lee and Lee 2011)

Table 1 Descriptive statistics of daily log-returns of NASDAQ and Nikkei 225

Statistics	NASDAQ	Nikkei 225
Mean	0.0007	0.0004
Standard deviation	0.0095	0.0130
Minimum	−0.0420	−0.0825
Median	0.0011	0.0006
Maximum	0.0415	0.0743
Kurtosis	5.2800	7.8141
Asymmetry	−0.6011	−0.1742

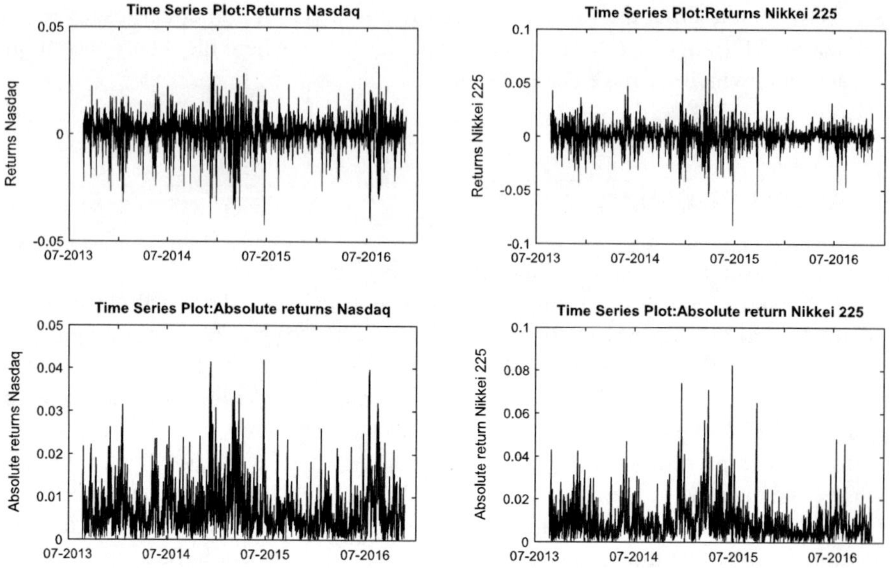

Fig. 1 Daily and absolute returns of NASDAQ and Nikkei 225

$$r_t = a_0 + \sum_{i=1}^{p} a_i r_{t-i} + \varepsilon_t + \sum_{j=1}^{q} b_j \varepsilon_{t-j}, \qquad \varepsilon_t = z_t \sigma_t,$$

$$\sigma_t^2 = c_0 + \sum_{i=1}^{r} c_i \varepsilon_{t-i}^2 + \sum_{j=1}^{s} d_j \sigma_{t-j}^2.$$

Here, z_t $(t = 1, 2, \ldots, T)$ is a sequence of independent and identically distributed (i.i.d.) random variables with K component Gaussian mixture density defined as

$$f_\eta(y) = \sum_{i=1}^{K} \pi_i f(y; \mu_i, \sigma_i),$$

where

$$f(y; \mu_i, \sigma_i) = \frac{1}{\sqrt{2\pi}\sigma_i} \exp\left\{ -\frac{1}{2} \left(\frac{y - \mu_i}{\sigma_i} \right)^2 \right\},$$

and the Gaussian mixture parameter is denoted by $\eta = (\pi_1, \ldots, \pi_K, \mu_1, \ldots, \mu_K, \sigma_1, \ldots, \sigma_K)$, and its space is

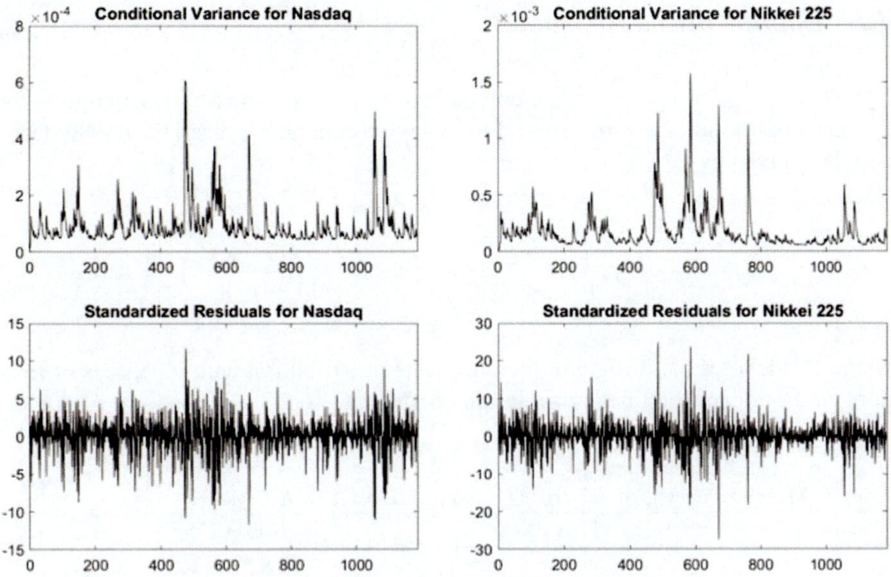

Fig. 2 Conditional variance and standardized residuals of NASDAQ and Nikkei 225

$$\Omega \subset \left\{ \eta \subset [0,1]^K \times \mathbb{R}^K \times (0,\infty)^K \mid \sum_{i=1}^{K} \pi_i = 1, \sum_{i=1}^{K} \pi_i \mu_i = 0, \sum_{i=1}^{K} \pi_i(\mu_i^2 + \sigma_i^2) = 1 \right\}.$$

For the estimation of parameters $\{a_i, b_j\}$, $\{c_i, d_j\}$, the so-called Gaussian quasi-maximum-likelihood estimation (QMLE) is utilized. See Lee and Lee (2011) for the detail. Observe also Palaro and Hotta (2006). The results for marginal models are given in Table 2. In fact, we have fitted two AR(1)-GARCH(1,1) for both series as initial models with a two-component Gaussian mixture. This selection was considered to see there was no autocorrelation nor squared autocorrelation in the residuals. Also it is usual to consider two or three components for the mixture of normals (here we only report the case of two). We also performed Ljung Box test Ljung and Box 1978 to infer that the null hypothesis is not rejected from lag 1 to 5. We report these values at Table 2. Finally, we report values for Kolmogorov–Smirnov (KS), Chi-square goodness-of-fit test (CSG), and Anderson–Darling test used for uniformity test for the standardized residuals (see Corder and Foreman 2014) (Figs. 2 and 3).

4.2 Copula-Based Approach

For a copula-based approach, the copulas we employ are the Student-t copula, the Plackett copula, and the symmetrized Joe-Clayton copula. We recall the the Student-t copula is given by

$$C^{\mathrm{t}}_{\rho,\nu}(u,v) = \int_{-\infty}^{\mathrm{t}_\nu^{-1}(u)} \int_{-\infty}^{\mathrm{t}_\nu^{-1}(v)} \frac{1}{2\pi\sqrt{1-\rho^2}} \left(1 + \frac{s^2 - 2\rho st + t^2}{\nu(1-\rho^2)}\right)^{-\frac{\nu+2}{2}} \mathrm{d}s\mathrm{d}t,$$

where t_ν^{-1} denotes the inverse of the univariate t distribution with ν degrees of freedom, and ρ means the linear correlation coefficient.

Table 2 Model parameters of univariate Gaussian mixture ARMA-GARCH

Parameter	NASDAQ	Nikkei 225
a_0	9.7400×10^{-4}	8.9296×10^{-4}
	(0.0059)	(0.0077)
a_1	−0.0340	−0.0333
	(0.8277)	(0.8046)
c_0	1.2871×10^{-5}	9.0378×10^{-6}
	(5.9042×10^{-5})	(4.9815×10^{-5})
c_1	0.1823	0.1600
	(0.7352)	(05737)
d_1	0.6747	0.7965
	(1.0831)	(0.6252)
π	0.7558	0.6788
	(1.9984)	(2.5686)
μ_1	0.1871	0.1258
	(1.2036)	(1.1685)
μ_2	−0.5793	−0.2658
	(6.0994)	(4.2117)
σ_1^2	0.5168	0.4309
	(1.8917)	(2.2835)
σ_2^2	2.0518	2.0984
	(8.8077)	(9.7640)
$Q^2(1)$	0.9906	0.7243
$Q^2(5)$	0.8659	0.5474
KS	0.2207	0.03806
χ^2	0.0888	0.0071
AD	0.0457	0.0083

Standard errors between brackets. Last values correspond to p-values for each test

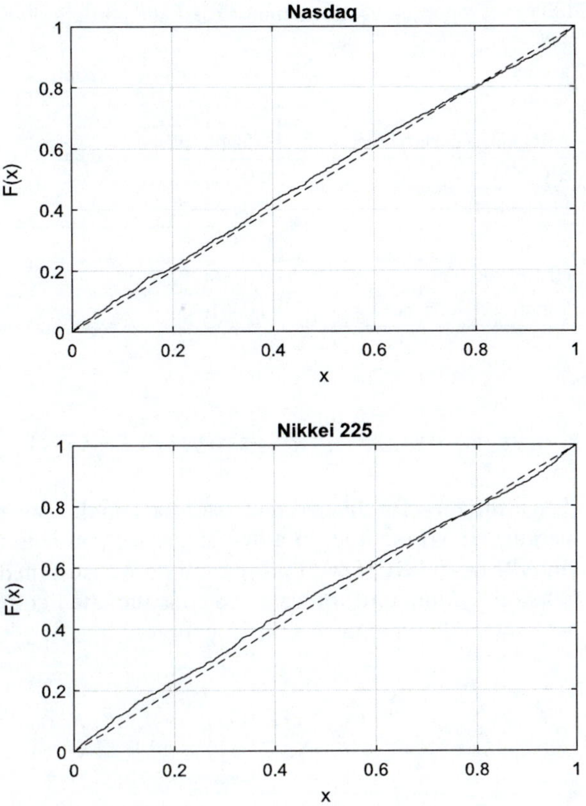

Fig. 3 Empirical distribution of transformed series u_t and v_t

The Plackett copula is given by

$$C_\theta(u, v) = \frac{1}{2(\theta-1)}\left\{1 + (\theta-1)(u+v) - \left((1+(\theta-1)(u+v))^2 - 4uv\theta(\theta-1)\right)^{\frac{1}{2}}\right\},$$

for $\theta > 0, \theta \neq 1$.

Finally, let λ_U, λ_L ($\in (0, 1)$) be two parameters, which will turn out to be the coefficient of upper and lower tail dependence, respectively (see Durante and Sempi 2016; Joe 1997; Nelsen 2006). The Joe-Clayton copula is given by

$$C_{\mathrm{JC}}(u, v; \lambda_U, \lambda_L) = 1 - \left(1 - \max\{(1-(1-u)^\kappa)^{-\gamma} + (1-(1-v)^\kappa)^{-\gamma} - 1, 0\}^{-\frac{1}{\gamma}}\right)^{\frac{1}{\kappa}},$$

where $\kappa = 1/\log_2(2-\lambda_U)$, $\gamma = -1/\log_2\lambda_L$. The symmetrized Joe-Clayton copula is now expressed as

Table 3 Proportion of observations where the portfolio loss exceeded the estimated VaR with copulas

Model	Proportion of violations	
	$\lambda = 0.05$	$\lambda = 0.01$
AR(1)-GARCH(1,1)	0.0644(45)	0.0200(14)
SJC-NM	0.0415(29)	0.0057(4)
Plackett-NM	0.0458(32)	0.0100(7)
t-student-NM	0.0443(31)	0.0086(6)
Historical	0.0386(27)	0.0086(6)
Variance–covariance	0.0601(42)	0.0229(16)

$$C_{\mathrm{SJC}}(u, v; \lambda_U, \lambda_L) = \frac{1}{2}C_{\mathrm{JC}}(u, v; \lambda_U, \lambda_L) + \frac{1}{2}C_{\mathrm{JC}}(1-u, 1-v; \lambda_L, \lambda_U) + u + v - 1.$$

The estimation for the parameters of the copula was made by inference function for margins (IFM) method. In a first stage, we compute the parameter for the margins via ARMA-GARCH with normal mixture distributed innovation. Once the data is transformed into uniform data, we construct the likelihood function and seek for the parameters that maximizes this function.

$$l(\theta) = \sum_{t=1}^{T} \ln c\left(F_1\left(r_{1t}; \theta_1\right), F_2\left(r_{2t}; \theta_2\right), \ldots, F_n\left(r_{nt}; \theta_n\right)\right) + \sum_{t=1}^{T}\sum_{j=1}^{n} \ln f_j\left(r_{jt}; \theta_j\right). \tag{4}$$

We again consider the portfolio of equal weight. First we estimate the parameters using the data from $t = 1$ to $t = 488$ as initial window and update the parameters each day as for the marginal distributions as for the copula. Our target is to find the solution of (3) for VaR at the level $\beta = 0.01$ and $\beta = 0.05$ concerning the data from $t = 489$ to $t = 1188$ (699 days). In Table 3, we can observe the proportion of observations where the loss exceeded confidence level. We then compare the forecast VaR with the actual return of the portfolio. However, the computation is highly demanding, and a Monte-Carlo simulation is preferred. Observing the value of violations, we could infer that the plain ARMA-GARCH with mixture of normal distributions is not well enough for this portfolio. But if we consider the effect of nonlinear dependence given by the copula, the improvement of implementation to the computation of VaR, we can see an outperform in both level of confidence. We also compared with benchmark models like variance–covariance and EWMA method. In all cases, the model with Plackett-Normal mixture gives the best results. Data is exhibited in Figs. 4 and 5.

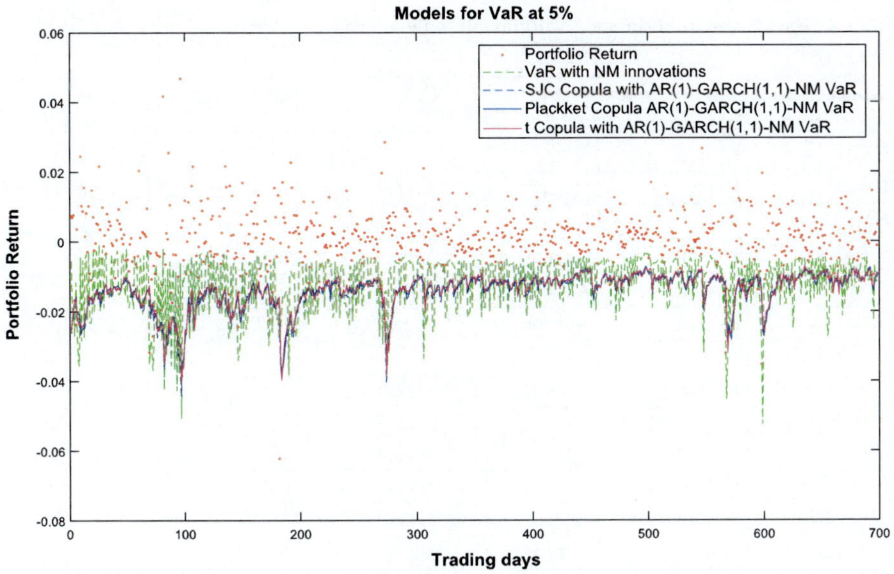

Fig. 4 One day ahead forecasts of VaR at $\beta = 5\%$ for portfolio of NASDAQ and Nikkei 225 with Gaussian mixture margins and various copulas

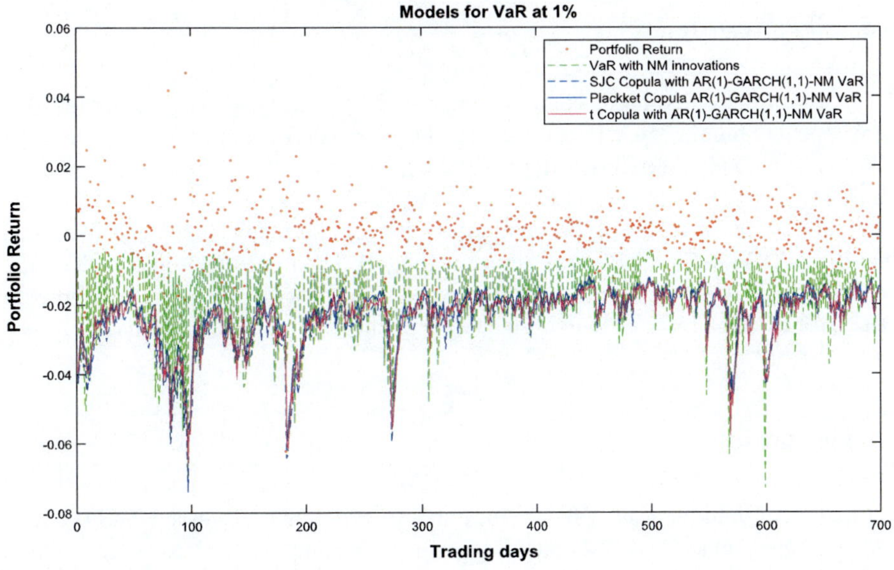

Fig. 5 One day ahead forecasts of VaR at $\beta = 1\%$ for portfolio of NASDAQ and Nikkei 225 with Gaussian mixture margins and various copulas

Table 4 Backtesting for estimated VaR models with copulas

Model	Bin	POF	CCI	Observations	Failures
$\beta = 0.05$					
AR-GARCH NM	Accept	Accept	Accept	699	45
Historical	Reject	Reject	Accept	699	27
Normal	Reject	Reject	Accept	699	42
SJC-NM	Accept	Accept	Accept	699	29
Plackett-NM	Accept	Accept	Accept	699	32
t-student-NM	Accept	Accept	Accept	699	31
$\beta = 0.01$					
AR-GARCH NM	Reject	Reject	Accept	699	14
Historical	Accept	Accept	Accept	699	6
Normal	Reject	Reject	Accept	699	16
SJC-NM	Accept	Accept	Accept	699	4
Plackett-NM	Accept	Accept	Accept	699	7
t-student-NM	Accept	Accept	Accept	699	6

4.3 Backtesting

To ascertain the outcome of computation, several backtestings are considered. We here appeal to binomial test (Bin), Kupiec's POF test (POF), and Christoffersen's test (CCI), respectively. See Christoffersen (1998), Kupiec (1995). The result is given in Table 4.

Again, we can infer the proposed models with copulas results in better estimations than plain ARMA-GARCH normal mixture models. Thanks to the property of copula, we can explain a better nonlinear correlation between the two indexes studied here. In effect, for extreme losses, copulas give better estimates and pass all the backtestings.

5 Discussions

Estimation of Value at Risk (VaR) for the portfolio problem is discussed. VaR is one of well used measures of risk. We consider the portfolio composed of Nasdaq and Nikkei 225 indexes, and estimate VaR empirically in two ways. One is a standard method which is based on ARMA-GARCH models with Gaussian mixture innovations; while the other is copula-based approach. Here we remark that a copula function is known to provide a flexible tool of handling nonlinear dependence between two indexes. In the evaluation of copula-based VaR, we appeal to the determination formula.

Implementation shows that compared to the former standard procedure, our copula-based outcome is indicated to be better.

There needs, however, more evidence to conclude that the copula-based approach is better in comparison with others. Our on-going research project is focused on empirical investigations for VaR estimation with various methods.

Acknowledgements We are grateful to an anonymous referee for precious comments, which helps in improving the manuscript. This work is partially supported by Grant-in-Aid for Scientific Research (Kakenhi) No. 15K04992 from Japan Society for the Promotion of Science (JSPS).

References

Christoffersen PF (1998) Evaluating interval forecasts. International Economic Review 39:841–862
Corder, G.W. Foreman, D.I.; Nonparametric Statistics: A Step-by-Step Approach. 2nd ed., Wiley (2014)
Duffie D, Pan J (1997) An overview of Value at Risk. J. Derivatives 4:7–49
Durante CF, Sempi C (2016) Principles of Copula Theory. CRC Press, Boca Raton
Fantazzini D (2008) Dynamic copula modelling for Value at Risk. Frontiers in Finance and Economics 5:72–108
Genest C, Favre AC (2007) Everything you always wanted to know about copula modeling but were afraid to ask. J. Hydrologic Engineering 12:347–368
Joe H (1997) Multivariate Models and Multivariate Dependence Concepts. Chapman and Hall/CRC Press, Boca Raton
Krzemienowski A, Szymczyk S (2016) Portfolio optimization with a coupla-based extension of conditional value-at-risk. Ann. Operations Research 237:219–236
Kupiec P (1995) Techniques for verifying the accuracy of risk measurement models. J. Derivatives 3:73–84
Lee S, Lee T (2011) Value at Risk forecasting based on Gaussian mixture ARMA-GARCH model. J. Statistical Computation and Simulation 81:1131–1144
Ljung GM, Box GEP (1978) On a measure of a lack of fit in time series models. Biometrika 65:297–303
McNeil AJ, Frey R, Embrechts P (2005) Quantitative Risk Management. Princeton University Press, Princeton
Molina Barreto, A.M. Ishimura, N. Yoshizawa, Y.; Value at risk for the portfolio problem with copulas, in "Empowering Science and Mathematics for Global Competitiveness," Y. Rahmawati and P.C. Taylor (Eds), CRC Press, London, 371–376 (2019)
Nelsen RB (2006) An Introduction to Copulas, 2nd edn. Springer Series in Statistics, Springer, New York
Palaro HP, Hotta LK (2006) Using conditional copula to estimate Value at Risk. J. Data Science 4:93–115
Prékopa A (2012) Multivariate value at risk and related topics. Ann. Operations Research 193:49–69
Sklar A (1973) Random variables, joint distribution functions, and copulas. Kybernetika 9:449–460

Notes on Backward Stochastic Differential Equations for Computing XVA

Jun Sekine and Akihiro Tanaka

1 Introduction

Backward stochastic differential equations (BSDEs) have been studied intensively from both theoretical and application viewpoints. Bismut (1976, 1978) studied BSDEs related to stochastic control problems, and Pardoux and Peng (1990) introduced general *nonlinear* BSDEs driven by Brownian motion as a noise process. After those early pioneering studies and since the late 1990s, the field of mathematical finance has provided various interesting research topics to develop the theory and application of BSDEs (e.g., El Karoui et al. 2000). In the present paper, we are interested in one such recent research topic in mathematical finance, namely the X-valuation adjustment (XVA) problem. The pricing and hedging methodology for over-the-counter (OTC) financial derivative securities for practitioners in financial institutions has been modified since the global financial crisis in 2008. The pre-crisis pricing was based on the Black–Scholes–Merton paradigm, and

$$p_{\mathrm{RN}} := \mathbb{E}\left[\mathrm{DF}_{\mathrm{r}}(T)\xi_T\right]$$

was regarded as the "fair" price of the derivative security (T, ξ_T). Here, ξ_T is a random variable representing the payoff at the maturity date $T \in \mathbb{R}_{++}(:= (0, \infty))$ of the derivative security, $\mathrm{DF}_{\mathrm{r}}(T) := \exp\left\{-\int_0^T r(u)\mathrm{d}u\right\}$ is a suitable discounting

This article is already available on arXiv:2006.02173v1 [q-fin.MF].
Jun Sekine's research is supported by a Grant-in-Aid for Scientific Research (C), No. 19K03636, from the Japan Society for the Promotion of Science.

J. Sekine (✉) · A. Tanaka
Graduate School of Engineering Science, Osaka University, Machikaneyama 1-3, Toyonaka 560-8531, Osaka, Japan
e-mail: sekine@sigmath.es.osaka-u.ac.jp

A. Tanaka
Sumitomo Mitsui Banking Corporation, 1-1-2, Marunouchi, Chiyoda-ku 100-0005, Tokyo, Japan

J. Cheng et al. (eds.), *Proceedings of the Forum "Math-for-Industry" 2018*,
Mathematics for Industry 35, https://doi.org/10.1007/978-981-16-5576-0_2

factor, where $r := (r(t))_{t\geq 0}$ is a risk-free interest rate process and $\mathbb{E}[(\cdot)]$ represents the expectation with respect to the so-called risk-neutral probability measure. By contrast, the post-crisis pricing formula used by practitioners in financial institutions is now described as

$$\bar{\mathrm{p}}_{\mathrm{RN}} + \sum_x x\mathrm{VA} \tag{1}$$

for the derivative security (T, ξ_T). Here,

$$\bar{\mathrm{p}}_{\mathrm{RN}} := \mathbb{E}\left[\mathrm{DF}_{\bar{r}}(T)\xi_T\right],$$

employing $\bar{r} := (\bar{r}(t))_{t\geq 0}$ as a risk-free interest rate process, which is different from r used in the pre-crisis model,[1] and

$$\sum_x x\mathrm{VA} = \mathrm{CVA} - \mathrm{DVA} + \mathrm{FVA} + \mathrm{ColVA} + \cdots$$

represents various valuation adjustments (e.g., credit valuation adjustment, debt valuation adjustment, funding valuation adjustment, collateral valuation adjustment). We may interpret the post-crisis modification as reflecting the following current situations.

1. The credit risk (default risk) of investors and their counterparties and the liquidity risk (of assets and cash) are widely recognized and and now considered seriously.
2. As a consequence of 1, the differences in various interest rates (e.g., risk-free rate, repo rate, funding rate, collateral rate) can no longer be neglected.

In this paper, we aim to understand the post-crisis pricing formula (1) in a better way from a theoretical viewpoint. Using BSDEs, which model the value processes of hedging portfolios, we interpret (1) as an approximate value of the fair price (i.e., the replication cost) of a derivative security. Concretely, this paper is organized as follows.

- In Sect. 2, we prepare a BSDE with a random horizon, where two random times τ_1, τ_2 and the progressively enlarged filtration by these random times are introduced, and the horizon is set as $\tau_1 \wedge \tau_2 \wedge T$ $(T \in \mathbb{R}_{++})$. We review some basic properties of such a BSDE, that is, the existence of a unique solution and its construction, using a reduced BSDE defined on a smaller filtration (see Theorems 1–3). These results are then used in Sect. 3.
- In Sect. 3, we construct a financial market model that generalizes the model given by Bichuch et al. (2018). On it, we derive BSDEs for pricing and hedging derivative securities, which express *nonlinear* dynamic hedging portfolio values of the seller and buyer. Here, we model the default time of the hedger (i.e., the seller of a

[1] The London Interbank Offered Rate (LIBOR) was a popular choice as the risk-free rate in pre-crisis models, whereas the Overnight Index Swap (OIS) rate is now recognized as a suitable candidate as the risk-free rate in post-crisis models.

derivative security) τ_1 and that of her counterparty (i.e., the buyer of the derivative security) τ_2, each of which are defined by random times. The contract between the hedger and her counterparty expires if the hedger or the counterparty defaults. Hence, $\tau_1 \wedge \tau_2 \wedge T$ is interpreted as the (random) horizon of the contract, where T is the prescribed fixed maturity, and we naturally have BSDEs considered in Sect. 2.

- In Sect. 4, working with the BSDEs introduced in Sect. 3, we obtain the following.

  1. An explicit sufficient condition is presented to ensure the non-existence of an arbitrage opportunity for both the seller and buyer of the derivative security (see Theorem 4). We note that a rather restrictive condition is necessary to ensure the existence of an arbitrage-free price (see Remark 14).
  2. The pricing formula (1) used by practitioners is interpreted as an approximation of the theoretical fair price of the derivative security: XVA is regarded as certain "zeroth" order approximated correction terms (see Theorem 5, Corollary 1, Proposition 3, and Remark 16). Furthermore, we mention a higher first-order approximation (see Sect. 4.3).

We intend to write this paper in an expository manner generally: Sect. 2 is devoted for reviewing known results and some results in Sect. 4 (that is, Theorem 4 and Propositions 1 and 2) are rather straightforward extensions of existing results of the closely related work by Bichuch et al. (2015, 2018) and Tanaka (2019). For other parts, we regard the following as being the contributions of the paper in comparison with Bichuch et al. (2015, 2018) and Tanaka (2019).

1. The market model is generalized: our model treats
   (a) a multiple risky asset model, and
   (b) a stochastic factor model that includes a stochastic volatility, a stochastic interest rate, and a stochastic hazard rate.
2. Different definitions of arbitrages and admissible trading strategies are employed (see Sect. 3.5). Because we analyze the pricing/hedging problem of derivative securities by using BSDEs, our choices seem to be natural and clear.
3. For XVA, an interpretation of pricing formula (1) is given as well as its arbitrage-free property (see Theorem 5, Corollary 1, and Proposition 3 with the following Remark 16 in Sect. 4.2, and cf. the results in Tanaka 2019).
4. Regarding the lending-borrowing spreads of interest rates as "small parameters," the first-order perturbed BSDEs are derived, and the associated approximated valuation adjustment terms are computed (see Proposition 4 in Sect. 4.3).

2 BSDE with a Random Horizon in a Progressively Enlarged Filtration

2.1 Setup

Let $(\Omega, \mathcal{F}, \mathbb{P})$ be a complete probability space, and let $W := (W(t))_{t\geq 0}$, $W(t) := (W_1(t), \dots, W_n(t))^\top$ be an n-dimensional Brownian motion on it. Define the filtration by

$$\mathcal{F}_t := \sigma\left(W(s);\, s \in [0, t]\right) \vee \mathcal{N}, \quad t \geq 0,$$

where $\mathcal{N}$ is the totality of null sets. Let E_1, E_2 be exponentially distributed random variables, assuming that W, E_1, and E_2 are mutually independent. Using nonnegative $\mathcal{F}_t$-progressively measurable processes $h_i := (h_i(t))_{t\geq 0}$, $(i = 1, 2)$, define the random times τ_1, τ_2 by

$$\tau_i := \inf\left\{ t \geq 0 \,\middle|\, \int_0^t h_i(u)\mathrm{d}u \geq E_i \right\}. \tag{2}$$

The indicator processes for τ_i $(i = 1, 2)$, namely

$$N_i(t) := 1_{\{t\geq \tau_i\}}, \quad t \geq 0,$$

are submartingales with respect to the filtration

$$\mathcal{H}_t := \sigma\left(N_1(s), N_2(s);\; s \in [0, t]\right), \quad t \geq 0,$$

and their Doob–Meyer decompositions are written as

$$N_i(t) = M_i(t) + \int_0^t \{1 - N_i(s)\}\, h_i(s)\mathrm{d}s, \quad t \geq 0$$

for $i = 1, 2$, where

$$M_i(t) := N_i(t) - \int_0^t \{1 - N_i(s)\}\, h_i(s)\mathrm{d}s, \quad t \geq 0$$

$(i = 1, 2)$ are two independent martingales with respect to $(\mathcal{H}_t)_{t\geq 0}$. Moreover, (W, M_1, M_2) remain as martingales with respect to the progressively enlarged filtration,

$$\mathcal{G}_t := \mathcal{F}_t \vee \mathcal{H}_t, \quad t \geq 0$$

(e.g., see Sect. 2.3 of Aksamit and Jeanblanc 2017), which are mutually independent. Also, we deduce that for $0 \le s \le t$,

$$\mathbb{P}\left(\tau_i > s \mid \mathcal{F}_t\right) = \mathbb{P}\left(\tau_i > s \mid \mathcal{F}_\infty\right) = \exp\left\{-\int_0^s h_i(u)\mathrm{d}u\right\},$$

where $\mathcal{F}_\infty := \sigma\left(\cup_{t\ge 0}\mathcal{F}_t\right)$. From this, we see that for $\mathrm{d}s \ll 1$,

$$\begin{aligned}\mathbb{P}\left(\tau_i \le s + \mathrm{d}s \mid \tau_i > s, \mathcal{F}_\infty\right) =& \frac{\mathbb{P}\left(s < \tau_i \le s + \mathrm{d}s|\mathcal{F}_\infty\right)}{\mathbb{P}\left(\tau_i > s|\mathcal{F}_\infty\right)}\\ =&1 - \exp\left\{-\int_s^{s+\mathrm{d}s} h_i(u)\mathrm{d}u\right\} \approx h_i(s)\mathrm{d}s,\end{aligned}$$

and h_i is called the hazard rate (or intensity) process for τ_i. Following Pham (2010), we employ the notation below.

Notation 1 • $\mathbb{F} := (\mathcal{F}_t)_{t\ge 0}$, $\mathbb{G} := (\mathcal{G}_t)_{t\ge 0}$, and $\mathbb{H} := (\mathcal{H}_t)_{t\ge 0}$.

- $\mathcal{P}(\mathbb{F})$ (resp. $\mathcal{P}(\mathbb{G})$): σ-algebra generated by $\mathbb{F}$ (resp. $\mathbb{G}$)-predictable measurable subsets on $\mathbb{R}_+ \times \Omega$. Equivalently, σ-algebra on $\mathbb{R}_+ \times \Omega$ generated by $\mathbb{F}$-adapted left-continuous processes.
- $\mathcal{O}(\mathbb{F})$ (resp. $\mathcal{O}(\mathbb{G})$): σ-algebra generated by $\mathbb{F}$ (resp. $\mathbb{G}$)-optional measurable subsets on $\mathbb{R}_+ \times \Omega$. Equivalently, σ-algebra on $\mathbb{R}_+ \times \Omega$ generated by $\mathbb{F}$-adapted right-continuous processes.
- $\mathcal{P}_\mathbb{F}$ (resp. $\mathcal{P}_\mathbb{G}$): the space of $\mathbb{F}$ (resp. $\mathbb{G}$)-predictable processes.
- $\mathcal{O}_\mathbb{F}$ (resp. $\mathcal{O}_\mathbb{G}$): the space of $\mathbb{F}$ (resp. $\mathbb{G}$)-optional processes.
- $\mathcal{P}_\mathbb{F}^{(k)}$: the space of the parametrized processes, $f : \mathbb{R}_+ \times \Omega \times \mathbb{R}_+^k \ni (t, \omega, u) \mapsto f_t(\omega, u) \in \mathbb{R}$, which is $\mathcal{P}(\mathbb{F}) \otimes \mathcal{B}(\mathbb{R}_+^k)/\mathcal{B}(\mathbb{R})$-measurable.
- $\mathcal{O}_\mathbb{F}^{(k)}$: the space of the parametrized processes, $f : \mathbb{R}_+ \times \Omega \times \mathbb{R}_+^k \ni (t, \omega, u) \mapsto f_t(\omega, u) \in \mathbb{R}$, which is $\mathcal{O}(\mathbb{F}) \otimes \mathcal{B}(\mathbb{R}_+^k)/\mathcal{B}(\mathbb{R})$-measurable.
- Denote by $\mathcal{P}_{\mathbb{F},t} := \left\{f 1_{[0,t]}|\ f \in \mathcal{P}_\mathbb{F}\right\}$, $\mathcal{O}_{\mathbb{F},t} := \left\{f 1_{[0,t]}|\ f \in \mathcal{O}_\mathbb{F}\right\}$, $\mathcal{P}_{\mathbb{F},t}^{(k)} := \left\{f(\cdot) 1_{[0,t]}|\ f \in \mathcal{P}_\mathbb{F}^{(k)}\right\}$, and $\mathcal{O}_{\mathbb{F},t}^{(k)} := \left\{f(\cdot) 1_{[0,t]}|\ f \in \mathcal{O}_\mathbb{F}^{(k)}\right\}$, for example.

We recall the following basic properties of stochastic processes under the progressively enlarged filtration $\mathbb{G}$.

Lemma 1 (Lemmas 5.1 and 2.1 of Pham 2010).

(1) Any $\mathcal{G}_t$-predictable process $(P(t))_{t\ge 0}$ has the expression that

$$\begin{aligned}P(t) =\ & p_0(t) 1_{\{t\le\tau_1\wedge\tau_2\}}\\ &+ p_t^1(\tau_1) 1_{\{\tau_1<t\le\tau_2\}} + p_t^2(\tau_2) 1_{\{\tau_2<t\le\tau_1\}} + p_t^{1,2}(\tau_1, \tau_2) 1_{\{t>\tau_1\vee\tau_2\}},\end{aligned}$$

where $(p_0(t))_{t\ge 0} \in \mathcal{P}_\mathbb{F}$, $\left(p_t^i(\cdot)\right)_{t\ge 0} \in \mathcal{P}_\mathbb{F}^{(1)}$ $(i = 1, 2)$ and $\left(p_t^{1,2}(\cdot, \cdot)\right)_{t\ge 0} \in \mathcal{P}_\mathbb{F}^{(2)}$.

(2) Any $\mathcal{G}_t$-optional process $(P(t))_{t\geq 0}$ has the expression that

$$P(t) = p_0(t)1_{\{t<\tau_1\wedge\tau_2\}} \\ +p_t^1(\tau_1)1_{\{\tau_1\leq t<\tau_2\}} + p_t^2(\tau_2)1_{\{\tau_2\leq t<\tau_1\}} + p_t^{1,2}(\tau_1,\tau_2)1_{\{t\geq\tau_1\vee\tau_2\}},$$

where $(p_0(t))_{t\geq 0} \in \mathcal{O}_{\mathbb{F}}$, $\left(p_t^i(\cdot)\right)_{t\geq 0} \in \mathcal{O}_{\mathbb{F}}^{(1)}$ $(i = 1, 2)$ and $\left(p_t^{1,2}(\cdot,\cdot)\right)_{t\geq 0} \in \mathcal{O}_{\mathbb{F}}^{(2)}$.

(3) Any $\mathcal{G}_t$-measurable random variable G_t has the expression that

$$G_t = g_t^0 1_{\{t<\tau_1\wedge\tau_2\}} \\ + g_t^1(\tau_1)1_{\{\tau_1\leq t<\tau_2\}} + g_t^2(\tau_2)1_{\{\tau_2\leq t<\tau_1\}} + g_t^{1,2}(\tau_1,\tau_2)1_{\{t\geq\tau_1\vee\tau_2\}},$$

where g_t^0 is an $\mathcal{F}_t$-measurable random variable, $\left(g_t^i(\cdot)\right)_{t\geq 0} \in \mathcal{O}_{\mathbb{F}}^{(1)}$ $(i = 1, 2)$, and $\left(g_t^{1,2}(\cdot,\cdot)\right)_{t\geq 0} \in \mathcal{O}_{\mathbb{F}}^{(2)}$.

Now, on the filtered probability space $(\Omega, \mathcal{F}, \mathbb{P}, \mathbb{G})$, we consider the BSDE

$$\begin{aligned} -\mathrm{d}Y(t) &= f(t, Y(t), Z(t), U_1(t), U_2(t))\,\mathrm{d}t \\ &\quad - Z(t)^\top \mathrm{d}W(t) - U_1(t)\mathrm{d}M_1(t) - U_2(t)\mathrm{d}M_2(t), \\ &\quad t \in [0, \tau_1\wedge\tau_2\wedge T], \\ Y(\tau_1\wedge\tau_2\wedge T) &= \phi_1(\tau_1)1_{\{\tau_1<\tau_2\wedge T\}} + \phi_2(\tau_2)1_{\{\tau_2<\tau_1\wedge T\}} + \xi_T 1_{\{T<\tau_1\wedge\tau_2\}}, \end{aligned} \tag{3}$$

where $T \in \mathbb{R}_{++} := (0, \infty)$ is a fixed terminal time, and the following conditions are imposed.

Assumption 1 (i) $\xi_T \in L^2(\Omega, \mathcal{F}_T, \mathbb{P})$.

(ii) For $i = 1, 2$, $\phi_i \in \mathcal{O}_{\mathbb{F}}$ so that $\mathbb{E}\left[\sup_{t\in[0,T]} |\phi_i(t)|^2\right] < \infty$.

(iii) $f : [0, T] \times \Omega \times \mathbb{R} \times \mathbb{R}^n \times \mathbb{R}^2 \to \mathbb{R}$ is $\mathcal{P}_{\mathbb{F}} \otimes \mathcal{B}(\mathbb{R}) \otimes \mathcal{B}(\mathbb{R}^n) \otimes \mathcal{B}(\mathbb{R}^2)/\mathcal{B}(\mathbb{R})$-measurable and satisfies, with some positive constant $K_{\mathrm{f}} > 0$,

$$\begin{aligned} &\left|f(t, y, z, u_1, u_2) - f\left(t, y', z', u_1', u_2'\right)\right| \\ &\leq K_{\mathrm{f}}\left(|y - y'| + |z - z'| + |u_1 - u_1'| + |u_2 - u_2'|\right) \\ &\text{for all}(y, z, u_1, u_2), (y', z', u_1', u_2') \end{aligned}$$

a.e. $(t, \omega) \in [0, T] \times \Omega$.

(iv) It holds that

$$\mathbb{E}\left[\int_0^T |f(t, 0, 0, 0, 0)|^2\,\mathrm{d}t\right] < \infty.$$

2.2 Existence, Uniqueness, and Construction of Solution

A specific feature of BSDE (3) is that it has the random time horizon $\tau_1 \wedge \tau_2 \wedge T$, where τ_i is the (first) jump time for the martingale M_i $(i = 1, 2)$. As for the definition of the solution to such a BSDE, we employ the following (cf. Darling and Pardoux 1997 as an example of related work).

Definition 1 We call the quadruplet $(Y, Z, U^1, U^2) : [0, T] \times \Omega \to \mathbb{R} \times \mathbb{R}^n \times \mathbb{R} \times \mathbb{R}$ a solution to BSDE (3) if it satisfies the following conditions.

(a) $Y := (Y(t))_{t\in[0,T]}$ is a $\mathbb{G}$-adapted RCLL (i.e., right continuous and having left limit) process (which is an element of $\mathcal{O}_{\mathbb{G},T}$), and $(Z, U^1, U^2) \in \left(\mathcal{P}_{\mathbb{G},T}\right)^{n+2}$.
(b) For $t \in [0, T]$, it holds that

$$
\begin{aligned}
Y(t)1_{\{\tau_1\wedge\tau_2\le t\}} &= \left\{\phi_1(\tau_1)1_{\{\tau_1<\tau_2\}} + \phi_2(\tau_2)1_{\{\tau_2<\tau_1\}}\right\} 1_{\{\tau_1\wedge\tau_2\le t\}},\\
Z(t)1_{\{\tau_1\wedge\tau_2\le t\}} &= 0,\\
U_i(t)1_{\{\tau_1\wedge\tau_2\le t\}} &= 0, \quad i = 1, 2.
\end{aligned}
$$

(c) For $t \in [0, T]$, it holds that

$$
\begin{aligned}
Y(t) = {} & \phi_1(\tau_1)1_{\{\tau_1<\tau_2,\tau_1\le T\}} + \phi_2(\tau_2)1_{\{\tau_2<\tau_1,\tau_2\le T\}} + \xi_T 1_{\{\tau_1\wedge\tau_2>T\}}\\
& + \int_{t\wedge\tau_1\wedge\tau_2}^{T\wedge\tau_1\wedge\tau_2} f(s, Y(s), Z(s), U_1(s), U_2(s))\,\mathrm{d}s\\
& - \int_{t\wedge\tau_1\wedge\tau_2}^{T\wedge\tau_1\wedge\tau_2} \left\{Z(s)^\top \mathrm{d}W(s) + U_1(s)\mathrm{d}M_1(s) + U_2(s)\mathrm{d}M_2(s)\right\}.
\end{aligned}
$$

Furthermore, we define the following spaces of stochastic processes, namely

$$
\begin{aligned}
\mathbb{S}^2_{\beta,T} &:= \left\{Y \in \mathcal{O}_{\mathbb{G},T} \mid \|Y\|^2_{\beta,T} < \infty\right\},\\
\mathbb{H}^{2,d}_{\beta,T} &:= \left\{Z \in \left(\mathcal{P}_{\mathbb{G},T}\right)^d \mid \|Z\|^2_{\beta,T} < \infty\right\},
\end{aligned}
$$

letting $\beta \in \mathbb{R}$ and denoting

$$
\|Y\|^2_{\beta,T} := \mathbb{E}\left[\int_0^T e^{\beta t}|Y(t)|^2 \mathrm{d}t\right].
$$

We then obtain the following.

Theorem 1 *Under Assumption 1, BSDE (3) admits a unique solution* $(Y, Z, U_1, U_2) \in \mathbb{S}^2_{\beta,T} \times \mathbb{H}^{2,n+2}_{\beta,T}$ *for any sufficiently large* $\beta > 0$.

Proof (Sketch). The method of proof is standard, although the horizon is random, which is rather "non-standard." We consider a Picard-type iteration, that is, for a given $\left(\bar{Y}, \bar{Z}, \bar{U}^1, \bar{U}^2\right) \in \mathbb{S}^2_{\beta,T} \times \mathbb{H}^{2,n+2}_{\beta,T}$, we construct the solution to BSDE

$$\begin{aligned}
-\mathrm{d}Y(t) &= f\left(t, \bar{Y}(t), \bar{Z}(t), \bar{U}_1(t), \bar{U}_2(t)\right)\mathrm{d}t \\
&\quad - Z(t)^\top dW(t) - U_1(t)\mathrm{d}M_1(t) - U_2(t)\mathrm{d}M_2(t), \\
&\quad t \in [0, \tau], \\
Y(\tau) &= \zeta,
\end{aligned} \tag{4}$$

where we denote

$$\begin{aligned}
&\tau_0 := \tau_1 \wedge \tau_2, \quad \tau := \tau_0 \wedge T, \\
&\zeta := \phi_1(\tau_1)1_{\{\tau_1 < \tau_2 \wedge T\}} + \phi_2(\tau_2)1_{\{\tau_2 < \tau_1 \wedge T\}} + \xi_T 1_{\{T < \tau_1 \wedge \tau_2\}}.
\end{aligned}$$

Indeed, using the $\mathbb{G}$-martingale representation

$$\begin{aligned}
\mathcal{M}(t) &:= \mathbb{E}\left[\zeta + \int_0^\tau f\left(u, \bar{Y}(u), \bar{Z}(u), \bar{U}_1(u), \bar{U}_2(u)\right)\mathrm{d}u \,\middle|\, \mathcal{G}_t\right] \\
&= \mathbb{E}\left[\zeta + \int_0^\tau f\left(u, \bar{Y}(u), \bar{Z}(u), \bar{U}_1(u), \bar{U}_2(u)\right)\mathrm{d}u\right] \\
&\quad + \int_0^t \phi(u)^\top \mathrm{d}W(u) + \int_0^t \psi_1(u)\mathrm{d}M_1(u) + \int_0^t \psi_2(u)dM_2(u), \quad t \in [0, T]
\end{aligned}$$

for some $(\phi, \psi^1, \psi^2) \in \mathbb{H}^{2,n+2}_{\beta,T}$ (e.g., see Sect. 5.2 of Bielecki and Rutkowski 2004), we define

$$\begin{aligned}
\tilde{Y}_t &:= \mathbb{E}\left[\zeta + \int_{t\wedge\tau}^{\tau} f\left(u, \bar{Y}_u, \bar{Z}_u, \bar{U}^1_u, \bar{U}^2_u\right)\mathrm{d}u \,\middle|\, \mathcal{G}_t\right], \quad t \in [0, T], \\
\tilde{Z} &:\equiv \phi, \quad \tilde{U}^1 :\equiv \psi^1, \quad \tilde{U}^2 :\equiv \psi^2.
\end{aligned}$$

Note that the martingale $(\mathcal{M}_t)_{t\in[0,T]}$ with respect to the right-continuous filtration $\mathbb{G}$ admits an RCLL modification. Hence,

$$\tilde{Y}(t) = \mathcal{M}(t) - \int_0^{t\wedge\tau} f\left(u, \bar{Y}(u), \bar{Z}(u), \bar{U}_1(u), \bar{U}_2(u)\right)\mathrm{d}u$$

also admits an RCLL modification, which is denoted by $\left(\tilde{Y}(t)\right)_{t\in[0,T]}$ again. Furthermore, we can check the integrability, $\tilde{Y} \in \mathbb{S}^2_{\beta,T}$. Hence, $\left(\tilde{Y}, \tilde{Z}, \tilde{U}_1, \tilde{U}_2\right)$ is the solution to (4). Next, we show that the map

$$\Psi : \mathbb{S}^2_{\beta,T} \times \mathbb{H}^{2,n+2}_{\beta,T} \ni (\bar{Y}, \bar{Z}, \bar{U}_1, \bar{U}_2) \mapsto \left(\tilde{Y}, \tilde{Z}, \tilde{U}_1, \tilde{U}_2\right) \in \mathbb{S}^2_{\beta,T} \times \mathbb{H}^{2,n+2}_{\beta,T}$$

is a contraction for sufficiently large $\beta > 0$, and using the fixed point theorem for the contraction map, we conclude that the fixed point of the map Ψ is the solution.

Remark 1 We refer to Sect. 19 of Cohen and Elliott (2015) for the detail of such a Picard-type iteration argument, where a more general semimartingale BSDE (driven by Lévy noise) is treated with a *fixed constant* time horizon.

Actually, we can construct the solution to BSDE (3) on the filtered probability space $(\Omega, \mathcal{F}, \mathbb{P}, \mathbb{G})$, using another reduced BSDE on the smaller filtered probability space $(\Omega, \mathcal{F}, \mathbb{P}, \mathbb{F})$. Assuming

Assumption 2 h_i $(i = 1, 2)$ are bounded,

we obtain the following.

Theorem 2 *Under Assumptions 1 and 2, the solution* $(Y, Z, U_1, U_2) \in \mathbb{S}^2_{\beta,T} \times \mathbb{H}^{2,n+2}_{\beta,T}$ *has the representation that*

$$\begin{aligned} Y(t) &= \bar{Y}(t)1_{\{0\le t<\tau_1\wedge\tau_2\wedge T\}} \\ &\quad + \left\{\phi_1(\tau_1)1_{\{\tau_1<\tau_2\wedge T\}} + \phi_2(\tau_2)1_{\{\tau_2<\tau_1\wedge T\}} + \xi_T 1_{\{T<\tau_1\wedge\tau_2\}}\right\}1_{\{t=\tau_1\wedge\tau_2\wedge T\}}, \\ Z(t) &= \bar{Z}(t), \\ U_i(t) &= \phi_i(t) - \bar{Y}(t), \quad i = 1, 2. \end{aligned} \tag{5}$$

Here, $(\bar{Y}, \bar{Z}) \in \mathbb{S}^2_{\beta,T} \times \mathbb{H}^{2,n}_{\beta,T}$ *is the solution to a BSDE on* $(\Omega, \mathcal{F}, \mathbb{P}, \mathbb{F})$*, namely*

$$\begin{aligned} -d\bar{Y}(t) &= \bar{f}\left(t, \bar{Y}(t), \bar{Z}(t)\right) dt - \bar{Z}(t)^\top dW(t), \quad t \in [0, T], \\ Y_T &= \xi_T, \end{aligned} \tag{6}$$

where

$$\begin{aligned} \bar{f}(t, y, z) := f\,(t, y, z, \phi_1(t) - y, \phi_2(t) - y) \\ + \{\phi_1(t) - y\} h_1(t) + \{\phi_2(t) - y\} h_2(t). \end{aligned}$$

Remark 2 Similar reduction results for BSDEs (into smaller filtrations) have been studied by Crépey and Song (2016) and Pham (2010) in more general settings.

Proof (Sketch). Note that BSDE (3) is rewritten as

$$
\begin{aligned}
-\mathrm{d}Y(t) &= \tilde{f}\,(t, Y(t), Z(t), U_1(t), U_2(t))\,\mathrm{d}t - Z(t)^\top \mathrm{d}W(t)\\
&\quad \text{on } \{0 \le t < \tau_1 \wedge \tau_2 \wedge T\},\\
\Delta Y(t) &= U_1(\tau_1)1_{\{\tau_1 < \tau_2 \wedge T\}} + U_2(\tau_2)1_{\{\tau_2 < \tau_1 \wedge T\}},\\
Y(t) &= \phi_1(\tau_1)1_{\{\tau_1 < \tau_2 \wedge T\}} + \phi_2(\tau_2)1_{\{\tau_2 < \tau_1 \wedge T\}} + F_T 1_{\{T < \tau_1 \wedge \tau_2\}}\\
&\quad \text{on } \{t = \tau_1 \wedge \tau_2 \wedge T\},
\end{aligned} \tag{7}
$$

where we use $\Delta Y(t) := Y(t) - Y(t-)$ and

$$
\tilde{f}\,(t, y, z, u_1, u_2) = f\,(t, y, z, u_1, u_2) + u_1 h_1(t) + u_2 h_2(t).
$$

We show that if we define (Y, Z, U^1, U^2) by (5), then it actually satisfies (7). First, we see that BSDE (6) on $(\Omega, \mathcal{F}, \mathbb{P}, \mathbb{F})$ has a unique solution $(\bar{Y}, \bar{Z}) \in \mathbb{S}^2_{\beta,T} \times \mathbb{H}^{2,n}_{\beta,T}$ for any sufficiently large $\beta > 0$, recalling that $\bar{f}$ is a standard driver (e.g., $\bar{f}(t, y, z)$ satisfies a globally Lipschitz condition with respect to (y, z)). Next, we can check that (5) indeed satisfies (7); for example, on $\{t = \tau_1 \wedge \tau_2 \wedge T\}$,

$$
\begin{aligned}
\Delta Y(t) &= \phi_1(\tau_1)1_{\{\tau_1 < \tau_2 \wedge T\}} + \phi_2(\tau_2)1_{\{\tau_2 < \tau_1 \wedge T\}} + \xi_T 1_{\{T < \tau_1 \wedge \tau_2\}} - \bar{Y}(t-)\\
&= \phi_1(\tau_1)1_{\{\tau_1 < \tau_2 \wedge T\}} + \phi_2(\tau_2)1_{\{\tau_2 < \tau_1 \wedge T\}} + \xi_T 1_{\{T < \tau_1 \wedge \tau_2\}}\\
&\quad - \left(\bar{Y}(\tau_1 \wedge \tau_2)1_{\{\tau_1 \wedge \tau_2 \le T\}} + \xi_T 1_{\{\tau_1 \wedge \tau_2 > T\}}\right)\\
&= U_1(\tau_1)1_{\{\tau_1 < \tau_2 \wedge T\}} + U_2(\tau_2)1_{\{\tau_2 < \tau_1 \wedge T\}}.
\end{aligned}
$$

Hence, the desired assertion follows as it is easy to see the integrabilities given by (5), $(Y, Z, U_1, U_2) \in \mathbb{S}^2_{\beta,T} \times \mathbb{H}^{2,n+2}_{\beta,T}$.

Remark 3 We impose Assumption 2 to simplify the statement of Theorem 2. We can relax it by employing a different solution space (from $\mathbb{S}^2_{\beta,T} \times \mathbb{H}^{2,n+2}_{\beta,T}$) associated with the so-called *stochastic* Lipschitz BSDEs. For the study of such BSDEs, see El Karoui and Huang (1997) and Nagayama (2019), for example.

2.3 Markovian Model

When we treat BSDE (3) in a practical application, more-concrete modeling is preferable: In this subsection, we consider BSDE (3) under Assumptions 1 and 2 and the following setting.

(i) There is a Markovian state variable process $X := (X(t))_{t \ge 0}$, which is governed by the following Markovian forward stochastic differential equation (FSDE), namely

$$
\mathrm{d}X(t) = b(t, X(t))\mathrm{d}t + a(t, X(t))\mathrm{d}W(t), \quad X(0) \in \mathbb{R}^d, \tag{8}
$$

on $(\Omega, \mathcal{F}, \mathbb{P}, (\mathcal{F}_t)_{t \ge 0})$, where $a : \mathbb{R}_+ \times \mathbb{R}^d \to \mathbb{R}^{d \times n}$ and $b : \mathbb{R}_+ \times \mathbb{R}^d \to \mathbb{R}^d$.

(ii) $h_i(t) := \tilde{h}_i(X(t))$, $i = 1, 2$, where $\tilde{h}_i : \mathbb{R}^d \to \mathbb{R}_+$ is bounded.

(iii) The driver $f : [0, T] \times \Omega \times \mathbb{R} \times \mathbb{R}^n \times \mathbb{R}^2 \to \mathbb{R}$ of BSDE (3) is written as

$$f(t, \omega, y, z, u_1, u_2) := g(t, X(t, \omega), y, z, u_1, u_2),$$

where $g : [0, T] \times \mathbb{R}^d \times \mathbb{R} \times \mathbb{R}^n \times \mathbb{R} \times \mathbb{R} \to \mathbb{R}$.

(iv) $\xi_T := \Xi(X(T))$, where $\Xi : \mathbb{R}^d \to \mathbb{R}$.

(v) $\phi_i(t) := \varphi_i(X(t)), i = 1, 2$, where $\varphi_i : \mathbb{R}^d \to \mathbb{R}$.

In this case, the solution to BSDE (3) can be constructed as follows using the solution to a second-order parabolic semilinear partial differential equation (PDE).

Theorem 3 *Consider the second-order parabolic semilinear PDE*

$$\begin{aligned} -\partial_t V(t, x) &= \mathcal{L}_t V(t, x) + \bar{g}\left(t, x, V(t, x), a(t, x)^\top \nabla V(t, x)\right), \\ &\qquad (t, x) \in [0, T) \times \mathbb{R}^d, \\ V(T, x) &= \Xi(x), \end{aligned} \tag{9}$$

where

$$\mathcal{L}_t V := \frac{1}{2} \mathrm{tr}\left(aa^\top(t, \cdot)\nabla\nabla V\right) + b^\top(t, \cdot)\nabla V \tag{10}$$

is the infinitesimal generator for X with the gradient $\nabla V := \left(\partial_{x_1} V, \dots, \partial_{x_d} V\right)^\top$ and the Hessian matrix $\nabla\nabla V := \left(\partial^2_{x_i x_j} V\right)_{1 \le i, j \le d}$, and

$$\bar{g}(t, x, y, z) := g\left(t, x, y, z, \varphi_1(x) - y, \varphi_2(x) - y\right) + \sum_{i=1}^{2} \{\varphi_i(x) - y\} \tilde{h}_i(x).$$

Suppose that there exists a unique classical solution $V \in C^{1,2}([0, T] \times \mathbb{R}^d)$ to (9). Then, the solution to BSDE (3) is represented as

$$\begin{aligned} Y(t) &= V\left(t, X(t)\right) 1_{\{0 \le t < \tau_1 \wedge \tau_2 \wedge T\}} + \Big\{\varphi_1\left(X(\tau_1)\right) 1_{\{\tau_1 < \tau_2 \wedge T\}} \\ &\quad + \varphi_2\left(X(\tau_2)\right) 1_{\{\tau_2 < \tau_1 \wedge T\}} + \Xi\left(X(T)\right) 1_{\{T < \tau_1 \wedge \tau_2\}}\Big\} 1_{\{t = \tau_1 \wedge \tau_2 \wedge T\}}, \\ Z(t) &= a\left(t, X(t)\right)^\top \nabla V\left(t, X(t)\right), \\ U_i(t) &= \varphi_i\left(X(t)\right) - V\left(t, X(t)\right), \quad i = 1, 2. \end{aligned}$$

Proof (Sketch). Associated with BSDE (6), we consider the (decoupled) forward-backward stochastic differential equation (FBSDE)

$$
\begin{aligned}
dX(t) &= b(t, X(t))\,\mathrm{d}t + a(t, X(t))\,\mathrm{d}W(t),\\
X(0) &\in \mathbb{R}^d,\\
-\mathrm{d}\bar{Y}(t) &= \bar{g}\left(t, X(t), \bar{Y}(t), \bar{Z}(t)\right)\mathrm{d}t - \bar{Z}(t)^\top \mathrm{d}W(t),\\
\bar{Y}(T) &= \Xi(X(T)).
\end{aligned}
\tag{11}
$$

By the nonlinear Feynman–Kac formula (e.g., see El Karoui et al. 2000 or Zhang 2017), the solution to (11) is expressed as

$$\bar{Y}(t) := V(t, X(t)), \quad \bar{Z}(t) := a(t, X(t))^\top \nabla V(t, X(t)), \quad t \in [0, T].$$

The desired assertion follows by using Theorem 2.

Remark 4 In the study of credit risk modeling in mathematical finance, similar techniques, namely the reduction of a BSDE (onto a Brownian filtration) combined with the (nonlinear) Feynman–Kac formula, have been utilized: see Bichuch et al. (2015), Bielecki et al. (2005), and Crépey (2015), for example.

3 XVA Calculation via BSDE

In this section, we introduce a "post-crisis" financial market model and a hedger's model for pricing OTC financial derivative securities, which generalize those employed by Bichuch et al. (2015), Bichuch et al. (2018), and Tanaka (2019). We then derive BSDEs that describe the self-financing hedging portfolio values of the hedger (seller) and her counterparty (buyer). After preparing mathematical models of a financial market, a hedger, and her counterparty, we formulate hedging problems and give the definition of the arbitrage-free price of a derivative security. Throughout this section, we continue to use the mathematical setup introduced in Sect. 2.

3.1 Non-defaultable/Defaultable Risky Assets

Let $T \in \mathbb{R}_{++}$ be a fixed time horizon, and consider a frictionless financial market model in continuous time. In it, there are price processes of n non-defaultable risky assets $S := (S_1, \ldots, S_n)^\top$, $S_i := (S_i(t))_{t\in[0,T]}$, one defaultable risky asset $P_{\mathrm{I}} := (P_{\mathrm{I}}(t))_{t\in[0,T]}$ issued by an investor's firm, and one defaultable risky asset $P_C := (P_C(t))_{t\in[0,T]}$ issued by the firm of a counterparty of the investor. They are governed by the following stochastic differential equations (SDEs) on $(\Omega, \mathcal{F}, \mathbb{P}, \mathbb{G})$:

$$\begin{aligned} \mathrm{d}S(t) =& \mathrm{diag}\left(S(t)\right)\left\{\sigma(t)\mathrm{d}W(t) + r_{\mathrm{D}}(t)\mathbf{1}\mathrm{d}t\right\}, \quad S(0) \in \mathbb{R}^n_{++}, && (12)\\ \mathrm{d}P_{\mathrm{I}}(t) =& P_{\mathrm{I}}(t-)\left\{\sigma_{\mathrm{I}}(t)\mathrm{d}W(t) - \mathrm{d}M_1(t) + r_{\mathrm{D}}(t)\mathrm{d}t\right\}, \quad P_{\mathrm{I}}(0) \in \mathbb{R}_{++}, && (13)\\ \mathrm{d}P_C(t) =& P_C(t-)\left\{\sigma_C(t)\mathrm{d}W(t) - \mathrm{d}M_2(t) + r_{\mathrm{D}}(t)\mathrm{d}t\right\}, \quad P_C(0) \in \mathbb{R}_{++}. && (14) \end{aligned}$$

Here, $\sigma \in (\mathcal{P}_{\mathbb{F},T})^{n\times n}$, $\sigma_i \in (\mathcal{P}_{\mathbb{F},T})^{1\times n}$, $i \in \{I, C\}$, and $r_{\mathrm{D}} \in \mathcal{P}_{\mathbb{F},T}$, which are assumed to be bounded, and $\sigma(t,\omega)$ is invertible for a.e. $(t,\omega) \in [0,T] \times \Omega$. Furthermore, we denote $\mathrm{diag}(x) = (x_i\delta_{ij})_{1\le i,j\le n}$ for $x := (x_1,\ldots,x_n)^\top \in \mathbb{R}^n$ and $\mathbf{1} := (1,\ldots,1)^\top \in \mathbb{R}^n$.

Remark 5 We regard the process r_{D} as the risk-free interest rate process in the market, which does not contain credit risk spread.[2] Define the cash account process $B_{\mathrm{D}} := (B_{\mathrm{D}}(t))_{t\ge 0}$ associated with the risk-free rate r_{D} by

$$\mathrm{d}B_{\mathrm{D}}(t) = B_{\mathrm{D}}(t) r_{\mathrm{D}}(t)\mathrm{d}t, \quad B_{\mathrm{D}}(0) = 1,$$

or equivalently

$$B_{\mathrm{D}}(t) = \exp\left\{\int_0^t r_{\mathrm{D}}(u)\mathrm{d}u\right\}.$$

We then see that

$$\frac{S_i}{B_{\mathrm{D}}}, \quad i = 1,\ldots,n, \quad \frac{P_j}{B_{\mathrm{D}}}, \quad j = 1,2$$

are $\mathbb{G}$-local martingales. These mean that we are starting with the probability space $(\Omega, \mathcal{F}, \mathbb{P})$ with a risk-neutral (pricing) probability $\mathbb{P}$,[3] not with the real-world (physical) probability.

The random times τ_1 and τ_2 defined by (2) are interpreted as the default times of the investor who issues P_{I} and the counterparty who issues P_C, respectively. We solve (13) as

$$\begin{aligned} P_{\mathrm{I}}(t) =& P_{\mathrm{I}}(0)\\ &\times \exp\left[\int_0^t \sigma_{\mathrm{I}}(u)\mathrm{d}W(u) + \int_0^t \left(r_{\mathrm{D}}(u) + h_1(u) - \frac{1}{2}|\sigma_{\mathrm{I}}(u)|^2\right)\mathrm{d}u\right]\{1 - N_1(t)\}, \end{aligned}$$

for example. Recall that the price becomes zero when defaults occur, i.e., $P_{\mathrm{I}}(\tau_1) = 0$.

Remark 6 As concrete examples of P_{I} and P_C, we can consider T-maturity zero coupon bonds without recoveries, namely

2 A typical example of such an interest rate in a real financial market is the OIS rate.

3 More precisely, $\mathbb{P}$ is an equivalent martingale measure (EMM). See Remark 13 in Sect. 3.5.

$$P_{\mathrm{I}}(t) = \mathbb{E}\left[\exp\left\{-\int_t^T (r_{\mathrm{D}}(u) + h_1(u))\,\mathrm{d}u\right\} \,\middle|\, \mathcal{F}_t\right]\{1 - N_1(t)\},$$

$$P_C(t) = \mathbb{E}\left[\exp\left\{-\int_t^T (r_{\mathrm{D}}(u) + h_2(u))\,\mathrm{d}u\right\} \,\middle|\, \mathcal{F}_t\right]\{1 - N_2(t)\}.$$

The volatility terms $(\sigma_j(t))_{t\in[0,T]}$ $(j \in \{I, C\})$ are described by using the $(\mathbb{P}, \mathcal{F}_t)$-Brownian martingale representation: For example, in the $j = I$ case, $(\sigma_{\mathrm{I}}(t))_{t\in[0,T]}$ is determined to satisfy

$$\begin{aligned}
&\mathbb{E}\left[\exp\left\{-\int_0^T (r_{\mathrm{D}}(u) + h_1(u))\,\mathrm{d}u\right\} \,\middle|\, \mathcal{F}_t\right] \\
&= P_{\mathrm{I}}(0)\exp\left\{\int_0^t \sigma_{\mathrm{I}}(s)\mathrm{d}W(s) - \frac{1}{2}\int_0^t |\sigma_{\mathrm{I}}(s)|^2\mathrm{d}s\right\} \quad \text{for} \quad t \in [0, T].
\end{aligned}$$

3.2 Defaultable Derivative Security

We treat the following derivative security in our financial market model.

Definition 2 A European derivative security is described as

$$(T, \tau_1, \tau_2, \xi_T, \phi_1, \phi_2),$$

where $\xi_T \in L^2(\Omega, \mathcal{F}_T, \mathbb{P})$ and $\phi_i \in \left\{\phi \in \mathcal{O}_{\mathbb{F},T} \,\middle|\, \mathbb{E}\left[\sup_{t\in[0,T]}|\phi(t)|^2\right] < \infty\right\}$ $(i = 1, 2)$. Here,

- $\tau_1 \wedge \tau_2 \wedge T$ is the maturity,
- ξ_T is the payoff at the maturity when no default occurs,
- $\phi_1(\tau_1)$ is the payoff at the maturity when the investor defaults,
- $\phi_2(\tau_2)$ is the payoff at the maturity when the counterparty defaults.

This means that at the maturity,

$$H := \xi_T 1_{\{T<\tau_1\wedge\tau_2\}} + \phi_1(\tau_1)1_{\{\tau_1<\tau_2,\tau_1\le T\}} + \phi_2(\tau_2)1_{\{\tau_2<\tau_1,\tau_2\le T\}} \tag{15}$$

is paid to the counterparty (buyer) from the investor (seller, writer).

Remark 7 A typical example of the payoff (ξ_T, ϕ_1, ϕ_2) is

$$\xi_T := h\left((S(t))_{t\in[0,T]}\right)$$

with $h : C([0,T], \mathbb{R}^n_{++}) \to \mathbb{R}$ and, for $i = 1, 2$,

$$\phi_i(t) := \varphi_i\left(\hat{V}(t)\right)$$

with some nonlinear (piecewise-linear) $\varphi_i : \mathbb{R} \to \mathbb{R}$ and

$$\hat{V}(t) := \mathbb{E}\left[\exp\left\{-\int_t^T r_{\mathrm{D}}(u)\mathrm{d}u\right\}\xi_T \,\middle|\, \mathcal{F}_t\right], \quad t \in [0,T]. \tag{16}$$

(16) is interpreted as the reference value process of the derivative (T, ξ_T) with the payoff ξ_T at the maturity T in a default-free market. In Bichuch et al. (2018),

$$\varphi_1(v) := v - L_{\mathrm{I}}(v - \alpha v)^+ \quad \text{and} \quad \varphi_2(v) := v + L_{\mathrm{C}}(v - \alpha v)^- \tag{17}$$

are employed, where $x^+ := \max(x, 0)$, $x^- := \max(-x, 0) = -\min(x, 0)$, $0 \le L_{\mathrm{I}}, L_{\mathrm{C}}, \alpha \le 1$. The constant L_{I} (resp. L_{C}) is called the loss rate upon default of the investor (resp. the counterparty), and α is called the collateralization level. For a more detailed explanation, see Sects. 3.2 and 3.4 of Bichuch et al. (2018).

3.3 Dynamic Portfolio Strategy

For hedging purposes, the writer (seller) of the derivative security given in Definition 2 constructs a dynamic portfolio, which is denoted by $\left(\pi, \pi^I, \pi^{\mathrm{C}}, \pi^{\mathrm{f}}, \pi^{\mathrm{r}}, \pi^{\mathrm{col}}\right)$. Here,

$$\pi := (\pi_1, \ldots, \pi_n)^\top \in \left(\mathcal{P}_{\mathbb{G},T}\right)^n, \quad \pi_j := (\pi_j(t))_{t\in[0,T]}$$

is an investment strategy for the risky assets $S := (S^1, \ldots, S^n)^\top$,

$$\pi^j := (\pi^j(t))_{t\in[0,T]} \in \mathcal{P}_{\mathbb{G},T}, \quad j \in \{I, C\}$$

are investment strategies for the risky assets P_{I} and P_{C}, respectively, and

$$\pi^j := (\pi^j(t))_{t\in[0,T]} \in \mathcal{P}_{\mathbb{G},T}, \quad j \in \{f, r, \mathrm{col}\}$$

are investment strategies for the cash accounts B_{f}, B_{r}, and B_{col}, which are called the funding account, the repo account, and the collateral account, respectively. They are defined by

$$\mathrm{d}B_j(t) = B_j(t)\left\{r_j^-(t)1_{\{\pi^j(t)<0\}} + r_j^+(t)1_{\{\pi^j(t)>0\}}\right\}\mathrm{d}t, \quad B_j(0) = 1 \tag{18}$$

with $r_j^- := (r_j^-(t))_{t\in[0,T]} \in \mathcal{P}_{\mathbb{F},T}$, $r_j^+ := (r_j^+(t))_{t\in[0,T]} \in \mathcal{P}_{\mathbb{F},T}$, and $j \in \{f, r, col\}$, where $r_{\mathrm{f}}^{\pm}$, $r_{\mathrm{r}}^{\pm}$ and $r_{\mathrm{col}}^{\pm}$ are called the funding rate, the repo rate, and the collateral rate, respectively.

Remark 8 The cash account process B_{f} represents the cumulative amount of cash that the hedger borrows from (or lends to) her treasury desk. The rate r_{f}^- is called the funding borrowing rate, and the rate r_{f}^+ is called the funding lending rate. The cash account process B_{r} represents the cumulative amount of cash that the investor borrows from (or lends to) a repo market. The rate r_{r}^- is called the repo borrowing rate, which is applied when the hedger borrows money from the repo market and implements a long position for the non-defaultable risky assets S. The rate r_{r}^+ is called the repo lending rate, which is applied when the hedger lends money to the repo market and implements a short-selling position for the non-defaultable risky assets S. The cash account process B_{col} represents the cumulative amount of cash that the investor receives from (or posts to) the counterparty as the collateral of the derivative security. The rate r_{col}^- is paid by the hedger to the counterparty if he/she has received the collateral. The rates r_{col}^+ is received by the hedger if he/she has posted the collateral. These rates can differ because different markets.[4] may be used to determine the contractual rates earned by cash collateral.

For $r_{\mathrm{f}}^{\pm}$ and $r_{\mathrm{r}}^{\pm}$, it is natural and realistic to assume that

$$2\epsilon_j :\equiv r_j^- - r_j^+ \geq 0 \quad \text{for } j \in \{f, r\}. \tag{19}$$

For $j \in \{f, r\}$, denoting the "mid-rate" by

$$r_j^0 :\equiv \frac{r_j^- + r_j^+}{2},$$

we see that

$$r_j^{\pm} \equiv r_j^0 \mp \epsilon_j.$$

The value process $Y := (Y(t))_{t\in[0,T]}$ associated with a given dynamic portfolio strategy $\left(\pi, \pi^{\mathrm{I}}, \pi^{\mathrm{C}}, \pi^{\mathrm{f}}, \pi^{\mathrm{r}}, \pi^{\mathrm{col}}\right)$ is governed by an SDE on $(\Omega, \mathcal{F}, \mathbb{P}, \mathbb{G})$, namely

$$\begin{aligned} \mathrm{d}Y(t) &= \pi(t)^{\top}\mathrm{d}S(t) + \pi^{\mathrm{I}}(t)\mathrm{d}P_{\mathrm{I}}(t) + \pi^{\mathrm{C}}(t)\mathrm{d}P_{\mathrm{C}}(t) \\ &\quad + \pi^{\mathrm{f}}(t)\mathrm{d}B_{\mathrm{f}}(t) + \pi^{\mathrm{r}}(t)\mathrm{d}B_{\mathrm{r}}(t) + \pi^{\mathrm{col}}(t)\mathrm{d}B_{\mathrm{col}}(t), \\ Y(0) &= y, \end{aligned} \tag{20}$$

subject to

[4] For example, the choice of currency (USD, Euro, etc.) We refer the interested reader to Fujii and Takahashi (2011), where the impact of the choice of currency of collateral is studied.

$$
\begin{aligned}
Y(t) &= \pi(t)^\top S(t) + \pi^{\mathrm{I}}(t)P_{\mathrm{I}}(t) + \pi^{\mathrm{I}}(t)P_{\mathrm{I}}(t) \\
&\quad + \pi^{\mathrm{f}}(t)B_{\mathrm{f}}(t) + \pi^{\mathrm{r}}(t)B_{\mathrm{r}}(t) + \pi^{\mathrm{col}}(t)B_{\mathrm{col}}(t), & (21) \\
\pi(t)^\top S(t) &+ \pi^{\mathrm{r}}(t)B_{\mathrm{r}}(t) = 0, & (22) \\
\pi^{\mathrm{col}}(t)B_{\mathrm{col}}(t) &- \alpha\hat{V}(t) = 0. & (23)
\end{aligned}
$$

Here, (21) corresponds to the so-called self-financing condition, (22) implies that the hedger accesses the repo market to purchase/sell non-defaultable risky assets (stocks), and (23) implies that $\alpha\hat{V}(t)$ is regarded as the collateral value at time t, where $\alpha \in [0, 1]$ is the collateral level, which is the same as the one given in Remark 7. From (21)–(23), recall that the relations

$$
\begin{aligned}
\pi^{\mathrm{r}}(t) &= -\,B_{\mathrm{r}}(t)^{-1}\pi(t)^\top S(t), & (24) \\
\pi^{\mathrm{col}}(t) &= B_{\mathrm{col}}(t)^{-1}\alpha\hat{V}(t), & (25) \\
\pi^{\mathrm{f}}(t) &= B_{\mathrm{f}}(t)^{-1}\left\{Y(t-) - \pi^{\mathrm{I}}(t)P_{\mathrm{I}}(t-) - \pi^{\mathrm{C}}(t)P_{\mathrm{C}}(t-) - \alpha\hat{V}(t)\right\} & (26)
\end{aligned}
$$

hold. Hence, we can interpret that $(y, \Pi) \in \mathbb{R} \times (\mathcal{P}_{\mathbb{G},T})^{n+2}$, where $\Pi := \left(\pi, \pi^I, \pi^{\mathrm{C}}\right)$, is a portfolio strategy that determines the portfolio value process (20), and we sometimes write

$$Y :\equiv Y^{(y,\Pi)},$$

emphasizing the portfolio strategy (y, Π). Combining (20) with (12)–(14), (18), and (24)–(26), we see that

$$
\begin{aligned}
\mathrm{d}Y(t) &= \pi(t)^\top \mathrm{diag}\,(S(t))\left[\sigma(t)\mathrm{d}W(t) + \left\{r_{\mathrm{D}}(t) - r_{\mathrm{r}}(t;\pi^{\mathrm{r}}(t))\right\}\mathbf{1}\mathrm{d}t\right] \\
&\quad + \pi^{\mathrm{I}}(t)P_{\mathrm{I}}(t-)\left[\sigma_{\mathrm{I}}(t)\mathrm{d}W(t) - \mathrm{d}M_1(t) + \left\{r_{\mathrm{D}}(t) - r_{\mathrm{f}}(t;\pi^{\mathrm{f}}(t))\right\}\mathrm{d}t\right] \\
&\quad + \pi^{\mathrm{C}}(t)P_{\mathrm{C}}(t-)\left[\sigma_{\mathrm{C}}(t)\mathrm{d}W(t) - \mathrm{d}M_2(t) + \left\{r_{\mathrm{D}}(t) - r_{\mathrm{f}}(t;\pi^{\mathrm{f}}(t))\right\}\mathrm{d}t\right] \\
&\quad + \left\{Y(t) - \alpha\hat{V}(t)\right\}r_{\mathrm{f}}(t;\pi^{\mathrm{f}}(t))\mathrm{d}t + \alpha\hat{V}(t)r_{\mathrm{col}}(t;\pi^{\mathrm{col}}(t))\mathrm{d}t, \qquad (27)
\end{aligned}
$$

where we denote

$$r_j(t;p) := r_j^-(t)1_{\{p<0\}} + r_j^+(t)1_{\{p>0\}}, \quad j \in \{f, r, col\}.$$

Remark 9 Suppose that $r_{\mathrm{D}} \equiv r_{\mathrm{f}}^{\pm} \equiv r_{\mathrm{r}}^{\pm} \equiv r_{\mathrm{col}}^{\pm}$. Then, (27) becomes

$$
\begin{aligned}
\mathrm{d}Y(t) &= \pi(t)^\top \mathrm{diag}\,(S(t))\,\sigma(t)\mathrm{d}W(t) + \pi^I(t)P_{\mathrm{I}}(t-)\left\{\sigma_{\mathrm{I}}(t)\mathrm{d}W(t) - \mathrm{d}M_1(t)\right\} \\
&\quad + \pi^{\mathrm{C}}(t)P_{\mathrm{C}}(t-)\left\{\sigma_{\mathrm{C}}(t)\mathrm{d}W(t) - \mathrm{d}M_2(t)\right\} + r_{\mathrm{D}}(t)Y(t)\mathrm{d}t,
\end{aligned}
$$

which is solved as

$$
\begin{aligned}
Y^{(y,\Pi)}(t) = B_{\mathrm{D}}(t)\Bigg[y &+ \int_0^t B_{\mathrm{D}}(s)^{-1}\pi(s)^{\top}\mathrm{diag}\,(S(s))\,\sigma(s)\mathrm{d}W(s) \\
&+ \int_0^t B_{\mathrm{D}}(s)^{-1}\pi^{\mathrm{I}}(s)P_{\mathrm{I}}(s-)\,\{\sigma_{\mathrm{I}}(s)\mathrm{d}W(s) - \mathrm{d}M_1(s)\} \\
&+ \int_0^t B_{\mathrm{D}}(s)^{-1}\pi^{\mathrm{C}}(s)P_{\mathrm{C}}(s-)\,\{\sigma_{\mathrm{C}}(s)\mathrm{d}W(s) - \mathrm{d}M_2(s)\}\Bigg]. \qquad (28)
\end{aligned}
$$

That is, the discounted value process Y/B_{D} is a local martingale, which is a standard result shared in a classical framework with "one risk-free rate world."

For the derivative security given in Definition 2, we call the portfolio strategy $(\hat{y}, \hat{\Pi}) \in \mathbb{R} \times (\mathcal{P}_{\mathbb{G},T})^{n+2}$ that satisfies

$$
Y^{(\hat{y},\hat{\Pi})}_{\tau_1\wedge\tau_2\wedge T} = H \qquad (29)
$$

the replicating portfolio strategy for the hedger.

Furthermore, for pricing purposes, we next consider a dynamic portfolio strategy $\left(-\tilde{\pi}, -\tilde{\pi}^{I}, -\tilde{\pi}^{\mathrm{C}}, \tilde{\pi}^{\mathrm{f}}, \tilde{\pi}^{\mathrm{r}}, \tilde{\pi}^{\mathrm{col}}\right)$ and the associated value process $\tilde{Y}$ of the buyer (counterparty). We define

$$
\begin{aligned}
\mathrm{d}\tilde{Y}(t) = &-\tilde{\pi}(t)^{\top}\mathrm{d}S(t) - \tilde{\pi}^{\mathrm{I}}(t)\mathrm{d}P_{\mathrm{I}}(t) - \tilde{\pi}^{\mathrm{C}}(t)\mathrm{d}P_{\mathrm{C}}(t) \\
&+\tilde{\pi}^{\mathrm{f}}(t)\mathrm{d}B_{\mathrm{f}}(t) + \tilde{\pi}^{\mathrm{r}}(t)\mathrm{d}B_{\mathrm{r}}(t) + \tilde{\pi}^{\mathrm{col}}(t)\mathrm{d}B_{\mathrm{col}}(t), \\
\tilde{Y}(0) = &-\tilde{y}
\end{aligned}
$$

subject to

$$
\begin{aligned}
\tilde{Y}(t) &= -\tilde{\pi}(t)^{\top}S(t) - \tilde{\pi}^{\mathrm{I}}(t)P_{I}(t) - \tilde{\pi}^{\mathrm{C}}(t)P_{C}(t) \\
&\qquad + \tilde{\pi}^{\mathrm{f}}(t)B_{\mathrm{f}}(t) + \tilde{\pi}^{\mathrm{r}}(t)B_{\mathrm{r}}(t) + \tilde{\pi}^{col}(t)B_{\mathrm{col}}(t), && (30) \\
&- \tilde{\pi}(t)^{\top}S(t) + \tilde{\pi}^{\mathrm{r}}(t)B_{\mathrm{r}}(t) = 0, && (31) \\
&\tilde{\pi}^{col}(t)B_{\mathrm{col}}(t) + \alpha\hat{V}(t) = 0, && (32)
\end{aligned}
$$

where $\tilde{\pi} \in \left(\mathcal{P}_{\mathbb{G},T}\right)^{n}$ and $\tilde{\pi}^{i} \in \mathcal{P}_{\mathbb{G},T}$ for $i \in \{I, C, f, r, col\}$. Here, as we see in (32), the collateral value at time t is regarded as $-\alpha\hat{V}(t)$, the opposite value of that for the writer (hedger). Because we see that

$$\begin{aligned}\tilde{\pi}^{\mathrm{r}}(t) =& B_{\mathrm{r}}(t)^{-1}\tilde{\pi}(t)^{\top}S(t),\\ \tilde{\pi}^{col}(t) =& -B_{\mathrm{col}}(t)^{-1}\alpha\hat{V}(t),\\ \tilde{\pi}^{\mathrm{f}}(t) =& B_{\mathrm{f}}(t)^{-1}\left\{\tilde{Y}(t-)+\tilde{\pi}^{\mathrm{I}}(t)P_I(t-)+\tilde{\pi}^{\mathrm{C}}(t)P_C(t-)+\alpha\hat{V}(t)\right\}\end{aligned}$$

from (30)–(32), we regard $\left(-\tilde{y}, -\tilde{\Pi}\right) \in \mathbb{R} \times \left(\mathcal{P}_{\mathbb{G},T}\right)^{n+2}$ with $\tilde{\Pi} := \left(\tilde{\pi}, \tilde{\pi}^{\mathrm{I}}, \tilde{\pi}^{\mathrm{C}}\right)$ as the portfolio strategy, and we rewrite the dynamics of $\tilde{Y} :\equiv \tilde{Y}^{(-\tilde{y},-\tilde{\Pi})}$ as

$$\begin{aligned}d\tilde{Y}(t) =& -\tilde{\pi}(t)^{\top}\mathrm{diag}\,(S(t))\left[\sigma(t)\mathrm{d}W(t)+\left\{r_{\mathrm{D}}(t)-r_{\mathrm{r}}(t;\pi^{\mathrm{r}}(t))\right\}\mathbf{1}\mathrm{d}t\right]\\ &-\tilde{\pi}^{\mathrm{I}}(t)P_I(t-)\left[\sigma_I(t)\mathrm{d}W(t)-\mathrm{d}M_1(t)+\left\{r_{\mathrm{D}}(t)-r_{\mathrm{f}}(t;\pi^{\mathrm{f}}(t))\right\}\mathrm{d}t\right]\\ &-\tilde{\pi}^{\mathrm{C}}(t)P_C(t-)\left[\sigma_C(t)\mathrm{d}W(t)-\mathrm{d}M_2(t)+\left\{r_{\mathrm{D}}(t)-r_{\mathrm{f}}(t;\pi^{\mathrm{f}}(t))\right\}\mathrm{d}t\right]\\ &+\left\{\tilde{Y}(t)+\alpha\hat{V}(t)\right\}r_{\mathrm{f}}(t;\pi^{\mathrm{f}}(t))\mathrm{d}t-\alpha\hat{V}(t)r_{\mathrm{col}}(t;\pi^{col}(t))\mathrm{d}t. \end{aligned} \quad (33)$$

Remark 10 We have assumed that the funding rate $r^{\pm}_{\mathrm{f,I}}$ for the investor (writer) and the funding rate $r^{\pm}_{\mathrm{f,C}}$ for the counterparty (buyer) are identical, i.e., $r^{\pm}_{\mathrm{f}} \equiv r^{\pm}_{f,I} \equiv r^{\pm}_{f,C}$, which is a restrictive situation. However, without such an assumption, it looks difficult and complicated to derive an explicit sufficient condition to ensure the no-arbitrage property (see Theorem 4 and its proof).

Remark 11 Suppose that $r_{\mathrm{D}} \equiv r^{\pm}_{\mathrm{f}} \equiv r^{\pm}_{\mathrm{r}} \equiv r^{\pm}_{\mathrm{col}}$. Using a similar calculation to that in Remark 9, we solve (33) to see that $\tilde{Y}^{(-y',-\tilde{\Pi})} \equiv -Y^{(y',\tilde{\Pi})}$, where the right-hand side $Y^{(y',\tilde{\Pi})}$ is given by (28) by letting $y := y'$ and $\Pi :\equiv \tilde{\Pi}$.

If the portfolio strategy $(-\tilde{y}, -\tilde{\Pi}) \in \mathbb{R} \times (\mathcal{P}_{\mathbb{G},T})^{n+2}$ satisfies

$$\tilde{Y}^{(-\tilde{y},-\tilde{\Pi})}_{\tau_1\wedge\tau_2\wedge T} = -H \quad (34)$$

for the derivative security given in Definition 2, then we call it the replicating portfolio strategy for the buyer.

3.4 Deriving BSDE

The replicating portfolio $(\hat{y}, \hat{\Pi})$ that satisfies (29) is represented using the solution to a BSDE. Let

$$\begin{aligned}Y^{+} :\equiv& Y^{(\hat{y},\hat{\Pi})},\\ U_1^{+}(t) :=& -\pi^{\mathrm{I}}(t)P_I(t-),\\ U_2^{+}(t) :=& -\pi^{\mathrm{C}}(t)P_C(t-),\\ Z^{+}(t) :=& \sigma(t)^{\top}\mathrm{diag}\,(S(t))\,\pi(t)-U_1^{+}(t)\sigma_I(t)^{\top}-U_2^{+}(t)\sigma_C(t)^{\top}.\end{aligned}$$

Recalling

$$\pi^{\mathrm{f}}(t)B_{\mathrm{f}}(t) = Y^{+}(t) + U_1^{+}(t) + U_2^{+}(t) - \alpha\hat{V}(t),$$

we see that $\pi^{\mathrm{f}}(t) \geq 0$ (resp. ≤ 0) is equivalent to

$$Y^{+}(t) + U_1^{+}(t) + U_2^{+}(t) - \alpha\hat{V}(t) \geq 0, \text{ (resp. } \leq 0).$$

Also, recalling

$$\begin{aligned}-\pi^{\mathrm{r}}(t)B_{\mathrm{r}}(t) &= \pi(t)^{\top}\mathrm{diag}(S(t))\mathbf{1}\\ &= \left\{Z^{+}(t)^{\top} + U_1^{+}(t)\sigma_I(t) + U_2^{+}(t)\sigma_C(t)\right\}\sigma(t)^{-1}\mathbf{1},\end{aligned}$$

we see that $\pi^{\mathrm{r}}(t) \geq 0$ (resp. ≤ 0) is equivalent to

$$\left\{Z^{+}(t)^{\top} + U_1^{+}(t)\sigma_I(t) + U_2^{+}(t)\sigma_C(t)\right\}\sigma(t)^{-1}\mathbf{1} \leq 0 \text{ (resp. } \geq 0).$$

Using these relations, we then rewrite (27) as

$$\begin{aligned}dY^{+}(t) = &-f^{+}\left(t, Y^{+}(t), Z^{+}(t), U_1^{+}(t), U_2^{+}(t); \hat{V}(t)\right)\mathrm{d}t\\ &+Z^{+}(t)^{\top}\mathrm{d}W(t) + U_1^{+}(t)\mathrm{d}M_1(t) + U_2^{+}(t)\mathrm{d}M_2(t),\end{aligned}$$

where

$$\begin{aligned}f^{+}\left(t, y, z, u_1, u_2; \hat{v}\right) := &f^{0}\left(t, y, z, u_1, u_2\right) + \alpha\left\{r_{\mathrm{f}}^{0}(t)\hat{v} - r_{\mathrm{col}}^{+}(t)\hat{v}^{+} + r_{\mathrm{col}}^{-}(t)\hat{v}^{-}\right\}\\ &+ \epsilon_{\mathrm{f}}(t)\left|y + u_1 + u_2 - \alpha\hat{v}\right|\\ &+ \epsilon_{\mathrm{r}}(t)\left|\left\{z^{\top} + u_1\sigma_I(t) + u_2\sigma_C(t)\right\}\sigma(t)^{-1}\mathbf{1}\right|, \end{aligned} \tag{35}$$

with

$$\begin{aligned}f^{0}\left(t, y, z, u_1, u_2\right) := &-r_{\mathrm{f}}^{0}(t)y + \left\{r_{\mathrm{r}}^{0}(t) - r_{\mathrm{D}}(t)\right\}z^{\top}\sigma(t)^{-1}\mathbf{1}\\ &+\left[-\left\{r_{\mathrm{f}}^{0}(t) - r_{\mathrm{D}}(t)\right\} + \left\{r_{\mathrm{r}}^{0}(t) - r_{\mathrm{D}}(t)\right\}\sigma_I(t)\sigma(t)^{-1}\mathbf{1}\right]u_1\\ &+\left[-\left\{r_{\mathrm{f}}^{0}(t) - r_{\mathrm{D}}(t)\right\} + \left\{r_{\mathrm{r}}^{0}(t) - r_{\mathrm{D}}(t)\right\}\sigma_C(t)\sigma(t)^{-1}\mathbf{1}\right]u_2.\end{aligned} \tag{36}$$

So, we consider the BSDE on the filtered probability space $(\Omega, \mathcal{F}, \mathbb{P}, \mathbb{G})$, namely

$$\begin{cases}-\mathrm{d}Y^{+}(t) = & f^{+}\left(t, Y^{+}(t), Z^{+}(t), U_1^{+}(t), U_2^{+}(t); \hat{V}(t)\right)\mathrm{d}t\\ & -Z^{+}(t)^{\top}\mathrm{d}W(t) - U_1^{+}(t)\mathrm{d}M_1(t) - U_2^{+}(t)\mathrm{d}M_2(t)\\ & \quad \text{for} \quad 0 \leq t \leq \tau_1 \wedge \tau_2 \wedge T,\\ Y^{+}\left(\tau_1 \wedge \tau_2 \wedge T\right) = H. & \end{cases} \tag{37}$$

Using the solution to (37), the replicating portfolio $\left(\hat{y}, \hat{\Pi}\right)$ that satisfies (29) is constructed as

$$\begin{aligned}
\hat{y} :=& Y^+(0),\\
\hat{\pi}(t) :=& \mathrm{diag}(S_t)^{-1}\left(\sigma(t)^\top\right)^{-1}\left\{Z^+(t) + U_1^+(t)\sigma_I^\top(t) + U_2^+(t)\sigma_C^\top(t)\right\},\\
\hat{\pi}^{\mathrm{I}}(t) :=& -P_I(t-)^{-1}U_1^+(t),\\
\hat{\pi}^{\mathrm{C}}(t) :=& -P_C(t-)^{-1}U_2^+(t)
\end{aligned}$$

for $0 \le t \le \tau_1 \wedge \tau_2 \wedge T$. Similarly, the replicating portfolio $(-\tilde{y}, -\tilde{\Pi})$ that satisfies (34) can be represented using the solution to a BSDE. Let

$$\begin{aligned}
Y^- :\equiv& -\tilde{Y}^{(-\tilde{y},-\tilde{\Pi})},\\
U_1^-(t) :=& -\tilde{\pi}^{\mathrm{I}}(t)P_{\mathrm{I}}(t-),\\
U_2^-(t) :=& -\tilde{\pi}^{\mathrm{C}}(t)P_{\mathrm{C}}(t-),\\
Z^-(t) :=& \sigma(t)^\top \mathrm{diag}\left(S(t)\right)\tilde{\pi}(t) - U_1^-(t)\sigma_I(t)^\top - U_2^-(t)\sigma_C(t)^\top.
\end{aligned}$$

Recalling

$$-\tilde{\pi}^{\mathrm{f}}(t)B_{\mathrm{f}}(t) = \tilde{Y}^-(t) + U_1^-(t) + U_2^-(t) - \alpha\hat{V}(t),$$

we see that $\pi^{\mathrm{f}}(t) \ge 0$ (resp. ≤ 0) is equivalent to

$$Y^-(t) + U_1^-(t) + U_2^-(t) - \alpha\hat{V}(t) \le 0, \text{ (resp. } \ge 0).$$

Also, recalling

$$\begin{aligned}
\tilde{\pi}^{\mathrm{r}}(t)B_{\mathrm{r}}(t) =& \tilde{\pi}(t)^\top \mathrm{diag}(S(t))\mathbf{1}\\
=& \left\{Z^-(t)^\top + U_1^-(t)\sigma_I(t) + U_2^-(t)\sigma_C(t)\right\}\sigma(t)^{-1}\mathbf{1},
\end{aligned}$$

we see that $\pi^{\mathrm{r}}(t) \ge 0$ (resp. ≤ 0) is equivalent to

$$\left\{Z^-(t)^\top + U_1^-(t)\sigma_I(t) + U_2^-(t)\sigma_C(t)\right\}\sigma(t)^{-1}\mathbf{1} \ge 0 \text{ (resp. } \le 0).$$

Using these relations, we then rewrite (33) as

$$\begin{aligned}
dY^-(t) =& -f^-\left(t, Y^-(t), Z^-(t), U_1^-(t), U_2^-(t); \hat{V}(t)\right)\mathrm{d}t\\
&+ Z^-(t)^\top \mathrm{d}W(t) + U_1^-(t)\mathrm{d}M_1(t) + U_2^-(t)\mathrm{d}M_2(t),
\end{aligned}$$

where

$$\begin{aligned} f^-\left(t,y,z,u_1,u_2;\hat{v}\right) :=& - f^+\left(t,-y,-z,-u_1,-u_2;-\hat{v}\right) \\ =& f^0\left(t,y,z,u_1,u_2\right) + \alpha\left\{r_{\mathrm{f}}^0(t)\hat{v} + r_{\mathrm{col}}^+(t)\hat{v}^- - r_{\mathrm{col}}^-(t)\hat{v}^+\right\} \\ & - \epsilon_{\mathrm{f}}(t)\left|y+u_1+u_2-\alpha\hat{v}\right| \\ & - \epsilon_{\mathrm{r}}(t)\left|\left\{z^\top + u_1\sigma_I(t) + u_2\sigma_C(t)\right\}\sigma(t)^{-1}\mathbf{1}\right|. \end{aligned} \tag{38}$$

So, we consider the BSDE on the filtered probability space $(\Omega, \mathcal{F}, \mathbb{P}, \mathbb{G})$

$$\begin{aligned} -\mathrm{d}Y^-(t) =& f^-\left(t, Y^-(t), Z^-(t), U_1^-(t), U_2^-(t); \hat{V}(t)\right)\mathrm{d}t \\ & - Z^-(t)^\top \mathrm{d}W(t) - U_1^-(t)\mathrm{d}M_1(t) - U_2^-(t)\mathrm{d}M_2(t) \\ & \text{for} \quad 0 \le t \le \tau_1 \wedge \tau_2 \wedge T, \\ Y^-\left(\tau_1 \wedge \tau_2 \wedge T\right) =& H. \end{aligned} \tag{39}$$

The replicating portfolio $\left(-\tilde{y}, -\tilde{\Pi}\right)$ that satisfies (34) is now constructed as

$$\begin{aligned} \tilde{y} :=& Y^-(0), \\ \tilde{\pi}(t) :=& \mathrm{diag}(S_t)^{-1}\left(\sigma(t)^\top\right)^{-1}\left\{Z^-(t) + U_1^-(t)\sigma_I^\top(t) + U_2^-(t)\sigma_C^\top(t)\right\}, \\ \tilde{\pi}^{\mathrm{I}}(t) :=& - P_I(t-)^{-1}U_1^-(t), \\ \tilde{\pi}^{\mathrm{C}}(t) :=& - P_C(t-)^{-1}U_2^-(t) \end{aligned}$$

for $0 \le t \le \tau_1 \wedge \tau_2 \wedge T$, using the solution to (39).

Remark 12 BSDEs (37) and (39) with (15) and (16) can be seen as the system of BSDEs

$$\begin{aligned} -\mathrm{d}Y^\pm(t) =& f^\pm\left(t, Y^\pm(t), Z^\pm(t), U_1^\pm(t), U_2^\pm(t); \hat{V}(t)\right)\mathrm{d}t \\ & - Z^\pm(t)^\top \mathrm{d}W(t) - U_1^\pm(t)\mathrm{d}M_1(t) - U_2^\pm(t)\mathrm{d}M_2(t), \\ & \text{for} \quad 0 \le t \le \tau_1 \wedge \tau_2 \wedge T, \\ Y^\pm\left(\tau_1 \wedge \tau_2 \wedge T\right) =& H, \\ -d\hat{V}(t) =& - r_{\mathrm{D}}(t)\hat{V}(t)\mathrm{d}t - \Delta(t)^\top \mathrm{d}W(t) \quad \text{for} \quad 0 \le t \le T, \\ \hat{V}(T) =& \xi_T, \end{aligned} \tag{40}$$

in which $\left(Y^\pm, Z^\pm, U_1^\pm, U_2^\pm, \hat{V}, \Delta\right)$ are solutions.

3.5 Hedging Problem

To study the hedging problem via BSDEs (37) and (39) with (15) and (16), it is natural to employ the following space of admissible hedging strategies

$$\mathscr{A}_{\beta,T} := \left\{ (\pi, \pi^{\mathrm{I}}, \pi^{\mathrm{C}}) \in (\mathcal{P}_{\mathbb{G},T})^{d+2} \;\middle|\; (\sigma^\top \mathrm{diag}(S)\pi, \pi^{\mathrm{I}} P_I^-, \pi^{\mathrm{C}} P_C^-) \in \mathbb{H}^{2,n+2}_{\beta,T} \right\},$$

where $\beta > 0$ is a fixed (sufficiently large) constant and we denote $P_i^-(t) := P_i(t-)$ for $t > 0$ and $P_i^-(0) := P_i(0)$. We then formulate the minimal superhedging price (i.e., the maximal price for the writer) and the maximal subhedging price (i.e., the minimal price for the buyer) as follows.

Definition 3 For the derivative security given in Definition 2,

$$\bar{p} := \inf \left\{ y \in \mathbb{R} \mid -H + Y^{(y,\Pi)}(\tau_1 \wedge \tau_2 \wedge T) \geq 0 \text{ for some } (y, \Pi) \in \mathbb{R} \times \mathscr{A}_{\beta,T} \right\}$$

is called the minimal superhedging price, which is the maximal price of the writer (seller), and

$$\underline{p} := \sup \left\{ y \in \mathbb{R} \mid H + \tilde{Y}^{(-y,-\Pi)}(\tau_1 \wedge \tau_2 \wedge T) \geq 0 \text{ for some } (y, \Pi) \in \mathbb{R} \times \mathscr{A}_{\beta,T} \right\}$$

is called the maximal subhedging price, which is the minimal price of the buyer. If there exists $\bar{\Pi} \in \mathscr{A}_{\beta,T}$ such that

$$-H + Y^{(\bar{p},\bar{\Pi})}(\tau_1 \wedge \tau_2 \wedge T) \geq 0,$$

then the pair $(\bar{p}, \bar{\Pi})$ is called the minimal superhedging strategy, and if there exists $\underline{\Pi} \in \mathscr{A}_{\beta,T}$ such that

$$H + \tilde{Y}^{(-\underline{p},-\underline{\Pi})}(\tau_1 \wedge \tau_2 \wedge T) \geq 0,$$

then the pair $\left(-\underline{p}, -\underline{\Pi}\right)$ is called the maximal subhedging strategy.

Associated with the hedging problem, we give the following definition.

Definition 4 Consider the derivative security given in Definition 2. Suppose that a writer sells the derivative security with price $p \in \mathbb{R}$ at time 0. If it holds that

$$-H + Y^{(p,\Pi)}(\tau_1 \wedge \tau_2 \wedge T) \geq 0 \quad \text{and} \quad \mathbb{P}\left(-H + Y^{(p,\Pi)}(\tau_1 \wedge \tau_2 \wedge T) > 0\right) > 0$$

for some $\Pi \in \mathscr{A}_{\beta,T}$, then we say that an arbitrage opportunity for the writer occurs. Similarly, suppose that a buyer purchases the derivative security with price $p \in \mathbb{R}$ at time 0. If it holds that

$$H + \tilde{Y}^{(-p,-\Pi)}(\tau_1 \wedge \tau_2 \wedge T) \geq 0 \quad \text{and} \quad \mathbb{P}\left(H + \tilde{Y}^{(-p,-\Pi)}(\tau_1 \wedge \tau_2 \wedge T) > 0\right) > 0$$

for some $\Pi \in \mathscr{A}_{\beta,T}$, then we say that an arbitrage opportunity for the buyer occurs. Moreover, if the price $\hat{p} \in \mathbb{R}$ at time 0 does not admit arbitrage opportunities for both writer and buyer, then $\hat{p}$ is called an arbitrage-free price.

Remark 13 In our financial market model, we assume implicitly that the probability measure $\mathbb{P}$ is an EMM. Hence, $\mathbb{P} \sim \mathbb{P}_0$, where $\mathbb{P}_0$ is a real-world (physical) probability measure given in the same measurable space $(\Omega, \mathcal{F})$. Therefore, in Definition 3, the $\mathbb{P}$-a.s. statement can be replaced by the $\mathbb{P}_0$-a.s. statement. Also, in Definition 4, $\mathbb{P}$ can be replaced by $\mathbb{P}_0$ to claim that $\mathbb{P}_0(\cdots) > 0$.

3.6 Markovian Model

The following Markovian model is typical and popularly treated in practice. Let the coefficients of the market model be described as

$$\begin{aligned} \sigma(t) :=& \tilde{\sigma}(t, F(t)), \quad r_{\mathrm{D}}(t) := \tilde{r}_{\mathrm{D}}(t, F(t)), \\ \sigma_i(t) :=& \tilde{\sigma}_j(t, F(t)), \quad i \in \{I, C\}, \\ h_j(t) :=& \tilde{h}_i(t, F(t)), \quad j \in \{1, 2\}, \\ r_k^0(t) :=& \tilde{r}_k^0(t, F(t)), \quad \epsilon_k(t) := \tilde{\epsilon}_k(t, F(t)), \quad k \in \{f, r\}, \\ \text{and} \quad r_{\mathrm{col}}^{\pm}(t) :=& \tilde{r}_{\mathrm{col}}^{\pm}(t, F(t)), \end{aligned}$$

where $\tilde{\sigma} : [0, T] \times \mathbb{R}^m \to \mathbb{R}^{n \times n}$, $\tilde{r}_{\mathrm{D}}, \tilde{\sigma}_i, \tilde{h}_j, \tilde{r}_k^0, \tilde{\epsilon}_k, \tilde{r}_{\mathrm{col}}^{\pm} : [0, T] \times \mathbb{R}^m \to \mathbb{R}$, and $(F(t))_{t \in [0,T]}$ is called the stochastic factor process, which can be interpreted as a model of economic factors and affects the market model through the coefficients $\sigma, \sigma_i\ (i \in \{I, C\}), h_j\ (j = 1, 2), r_k^0, \epsilon_k^0\ (k \in \{f, r\})$, and $r_{\mathrm{col}}^{\pm}$. It is given by the solution to the SDE

$$dF(t) = \mu_F(t, F(t))\, \mathrm{d}t + \sigma_F(t, F(t))\, \mathrm{d}W(t), \quad F(0) \in \mathbb{R}^m$$

on $(\Omega, \mathcal{F}, \mathbb{P}, \mathbb{F})$, where $\mu_F : [0, T] \times \mathbb{R}^m \to \mathbb{R}^m$ and $\sigma_F : [0, T] \times \mathbb{R}^m \to \mathbb{R}^{m \times n}$. Let

$$X^\top :\equiv \left(X_1^\top, X_2^\top\right) :\equiv \left(S^\top, F^\top\right)$$

and define, for $x := (x_1, x_2) \in \mathbb{R}^n \times \mathbb{R}^m$,

$$b(t, x) := \begin{pmatrix} \mathrm{diag}(x_1) r_{\mathrm{D}}(t, x_2) \\ \mu_F(t, x_2) \end{pmatrix}, \quad a(t, x) := \begin{pmatrix} \mathrm{diag}(x_1) \sigma(t, x_2) \\ \sigma_F(t, x_2) \end{pmatrix}.$$

Then, the SDE for X is written as (8) with $d = n + m$. Furthermore, we set

$$\xi_T := \Xi(X(T)) \quad \text{and} \quad \phi_i(t) := \varphi_i(\hat{V}(t)) \quad \text{for } i \in \{1, 2\},$$

where $\Xi : \mathbb{R}^{n+m} \to \mathbb{R}$ and $\varphi_i : \mathbb{R} \to \mathbb{R}$. In this situation, we can apply Theorem 3 to represent the solution to BSDEs (40), using the solutions to the associated PDEs (see Proposition 2 in Sect. 4).

4 Results

Throughout this section, we always assume that σ_i ($i \in \{I, C\}$), $\sigma, \sigma^{-1}, r_{\mathrm{D}}, r_j^{\pm}$ ($j \in \{f, r, col\}$), and h_k ($k = 1, 2$) are bounded. Applying the results in Sect. 2 and a comparison theorem for BSDEs, the following claims are straightforward to see.

Proposition 1 *For any sufficiently large* $\beta > 0$, *there exist unique solutions* $\left(Y^{\pm}, Z^{\pm}, U_1^{\pm}, U_2^{\pm}\right) \in \mathbb{S}^2_{\beta,T} \times \mathbb{H}^{2,n+2}_{\beta,T}$ *to BSDEs (37) and (39) with (15) and (16). Moreover, the solutions have the representations that*

$$
\begin{aligned}
Y^{\pm}(t) &= \bar{Y}^{\pm}(t) 1_{\{0 \le t < \tau_1 \wedge \tau_2 \wedge T\}} \\
&+ \left\{\phi_1(\tau_1) 1_{\{\tau_1 < \tau_2 \wedge T\}} + \phi_2(\tau_2) 1_{\{\tau_2 < \tau_1 \wedge T\}} + \xi_T 1_{\{T < \tau_1 \wedge \tau_2\}}\right\} 1_{\{t = \tau_1 \wedge \tau_2 \wedge T\}}, \\
Z^{\pm}(t) &= \bar{Z}^{\pm}(t), \\
U_i^{\pm}(t) &= \phi_i(t) - \bar{Y}^{\pm}(t), \quad i = 1, 2.
\end{aligned} \tag{41}
$$

Here, $\left(\bar{Y}^{\pm}, \bar{Z}^{\pm}\right) \in \mathbb{S}^2_{\beta,T} \times \mathbb{H}^{2,n}_{\beta,T}$ *are the solutions to BSDEs on* $(\Omega, \mathcal{F}, \mathbb{P}, \mathbb{F})$, *namely*

$$
\begin{aligned}
-d\bar{Y}^{\pm}(t) &= \bar{f}^{\pm}\left(t, \bar{Y}^{\pm}(t), \bar{Z}^{\pm}(t); \hat{V}(t), \phi_1(t), \phi_2(t)\right) dt - \bar{Z}^{\pm}(t)^{\top} dW(t) \\
&\qquad \textit{for } \ 0 \le t \le T, \\
\bar{Y}^{\pm}(T) &= \ \xi_T, \\
-d\hat{V}(t) &= \ -r_D(t)\hat{V}(t) dt - \Delta(t)^{\top} dW(t) \ \textit{ for } \ 0 \le t \le T, \\
\hat{V}(T) &= \ \xi_T,
\end{aligned} \tag{42}
$$

where we define

$$
\begin{aligned}
&\bar{f}^{\pm}\left(t, y, z; \hat{v}, p_1, p_2\right) \\
&\quad := f^{\pm}\left(t, y, z, p_1 - y, p_2 - y; \hat{v}\right) + (p_1 - y) h_1(t) + (p_2 - y) h_2(t).
\end{aligned} \tag{43}
$$

In addition to Condition (19), assume that

$$
r_{col}^{-} \ge r_{col}^{+}. \tag{44}
$$

Then, it always holds that

$$
Y^{-} \le Y^{+} \ \textit{ and } \ \bar{Y}^{-} \le \bar{Y}^{+}. \tag{45}
$$

Proof (Sketch). Using (19) and (44), we see that

$$\begin{aligned}&\bar{f}^{+}\left(t,y,z;\hat{v},p_1,p_2\right)-\bar{f}^{-}\left(t,y,z;\hat{v},p_1,p_2\right)\\ =&\alpha\left\{r^{-}_{\mathrm{col}}(t)-r^{+}_{\mathrm{col}}(t)\right\}|\hat{v}|+2\epsilon_{\mathrm{f}}(t)\left|y+(p_1-y)+(p_2-y)-\alpha\hat{v}\right|\\ &+2\epsilon_{\mathrm{r}}(t)\left|\left\{z^{\top}+(p_1-y)\sigma_I(t)+(p_2-y)\sigma_C(t)\right\}\sigma(t)^{-1}\mathbf{1}\right|\geq 0.\end{aligned}$$

Hence, (45) follows from a comparison theorem of BSDEs. Other assertions follow from the results in Sect. 2.

Next, consider the Markovian model given in Sect. 3.6. Then, corresponding to (42), we have the Markovian system of BSDEs (decoupled FBSDEs)

$$\begin{aligned}dX(t)&=b(t,X(t))\mathrm{d}t+a(t,X(t))\mathrm{d}W(t),\quad X(0)\in\mathbb{R}^{n+m},\\ -\mathrm{d}\bar{Y}^{\pm}(t)&=\bar{g}^{\pm}\left(t,X_2(t),\bar{Y}^{\pm}(t),\bar{Z}^{\pm}(t);\hat{V}(t),\varphi_1(\hat{V}(t)),\varphi_2(\hat{V}(t))\right)\mathrm{d}t\\ &\quad-\bar{Z}^{\pm}(t)^{\top}\mathrm{d}W(t),\\ \bar{Y}^{\pm}(T)&=\Xi\left(X(T)\right),\\ -d\hat{V}(t)&=-\tilde{r}_{\mathrm{D}}(t,X_2(t))\hat{V}(t)\mathrm{d}t-\Delta(t)^{\top}\mathrm{d}W(t),\\ \hat{V}(T)&=\Xi\left(X(T)\right).\end{aligned}\tag{46}$$

Here, the relation

$$\bar{g}^{\pm}(t,X_2(t,\omega),y,z;\hat{v},p_1,p_2)=\bar{f}^{\pm}(t,\omega,y,z;\hat{v},p_1,p_2)$$

holds, and the functions $\bar{g}^{\pm}:[0,T]\times\mathbb{R}^m\times\mathbb{R}\times\mathbb{R}^n\times\mathbb{R}^3\to\mathbb{R}$ are written as

$$\begin{aligned}&\bar{g}^{\pm}(t,x_2,y,z;\hat{v},p_1,p_2):=\bar{g}^{0}(t,x_2,y,z;p_1,p_2)\\ &+\alpha\left\{\tilde{r}^{0}_{\mathrm{f}}(t,x_2)\hat{v}\mp\tilde{r}^{+}_{\mathrm{col}}(t,x_2)\hat{v}^{\pm}\pm\tilde{r}^{-}_{\mathrm{col}}(t,x_2)\hat{v}^{\mp}\right\}\\ &\pm\tilde{\epsilon}_{\mathrm{f}}(t,x_2)\left|y+(p_1-y)+(p_2-y)-\alpha\hat{v}\right|\\ &\pm\tilde{\epsilon}_{\mathrm{r}}(t,x_2)\left|\left\{z^{\top}+(p_1-y)\tilde{\sigma}_I(t,x_2)+(p_2-y)\tilde{\sigma}_C(t,x_2)\right\}\tilde{\sigma}(t,x_2)^{-1}\mathbf{1}\right|\end{aligned}$$

with

$$\begin{aligned}&\bar{g}^{0}(t,x_2,y,z;p_1,p_2):=z^{\top}\left\{(\tilde{r}^{0}_{\mathrm{r}}-\tilde{r}_{\mathrm{D}})\tilde{\sigma}^{-1}\mathbf{1}\right\}(t,x_2)\\ &-\left\{(2\tilde{r}_{\mathrm{D}}-\tilde{r}^{0}_{\mathrm{f}}+\tilde{h}_1+\tilde{h}_2)+(\tilde{r}^{0}_{\mathrm{r}}-\tilde{r}_{\mathrm{D}})(\tilde{\sigma}_I+\tilde{\sigma}_C)\tilde{\sigma}^{-1}\mathbf{1}\right\}(t,x_2)y\\ &+\left\{\tilde{h}_1-(\tilde{r}^{0}_{\mathrm{f}}-\tilde{r}_{\mathrm{D}})+(\tilde{r}^{0}_{\mathrm{r}}-\tilde{r}_{\mathrm{D}})\tilde{\sigma}_I\tilde{\sigma}^{-1}\mathbf{1}\right\}(t,x_2)p_1\\ &+\left\{\tilde{h}_2-(\tilde{r}^{0}_{\mathrm{f}}-\tilde{r}_{\mathrm{D}})+(\tilde{r}^{0}_{\mathrm{r}}-\tilde{r}_{\mathrm{D}})\tilde{\sigma}_C\tilde{\sigma}^{-1}\mathbf{1}\right\}(t,x_2)p_2.\end{aligned}$$

Utilizing Theorem 3, we obtain the following.

Proposition 2 *Denote* $d := n + m$ *and consider the system of second-order parabolic semilinear PDEs*

$$\begin{aligned}
-\partial_t V &= \{\mathcal{L}_t - \tilde{r}_D(t, x_2)\} V, \quad (t, x) \in [0, T) \times \mathbb{R}^d, \\
V(T, x) &= \Xi(x), \\
-\partial_t U^{\pm} &= \mathcal{L}_t U^{\pm} + \tilde{g}^{\pm}\big(t, x_2, U, a^\top \nabla U^{\pm}; V, \varphi_1(V), \varphi_2(V)\big), \\
&\quad (t, x) \in [0, T) \times \mathbb{R}^d, \\
U^{\pm}(T, x) &= \Xi(x),
\end{aligned} \tag{47}$$

where $\mathcal{L}_t(\cdot)$ *is the infinitesimal generator for* X *given by (10). Suppose that there exists a unique classical solution* $\big(V, U^{\pm}\big) \in \big(C^{1,2}([0, T] \times \mathbb{R}^d)\big)^2$ *to (47). Then the solution to BSDE (46) is represented as*

$$\bar{Y}^{\pm}(t) = U^{\pm}(t, X(t)), \quad \bar{Z}^{\pm}(t) = \big(a \nabla U^{\pm}\big)(t, X(t)), \quad t \in [0, T].$$

4.1 Results on Arbitrage

Theorem 4 *In addition to Conditions (19) and (44), assume the following:*

$$\begin{aligned}
h_1 &\geq r_f^- - r_D - \big(r_r^+ - r_D\big)(\sigma_I \sigma^{-1}\mathbf{1})^+ + \big(r_r^- - r_D\big)(\sigma_I \sigma^{-1}\mathbf{1})^-, \\
h_2 &\geq r_f^- - r_D - \big(r_r^+ - r_D\big)(\sigma_C \sigma^{-1}\mathbf{1})^+ + \big(r_r^- - r_D\big)(\sigma_C \sigma^{-1}\mathbf{1})^-,
\end{aligned} \tag{48}$$

and

$$r_f^+ \geq r_{col}^-. \tag{49}$$

Then, it holds that $\underline{p} = Y^-(0) \leq Y^+(0) = \bar{p}$. *Hence, for the derivative security given in Definition 2, any price* $p \in \big[Y^-(0), Y^+(0)\big]$ *at time* 0 *is arbitrage-free.*

Remark 14 The conditions imposed in Theorem 4 to ensure the arbitrage-free property look to be rather strong: violating (44), (48), or (49) seems to be realizable in real situations. Relaxing the arbitrage-free condition by admitting "certain" arbitrage opportunities might be an interesting research direction for this bilateral hedging scheme with collateralizations. We refer to Thoednithi (2015) and Nie and Rutkowski (2018) as related studies.

Proof (Sketch). Using (35), (36), (38), and (43), we see that

$$\begin{aligned}
&\bar{f}^{\pm}\left(t, y, z; \hat{v}, p_1, p_2\right) \\
=&z^{\top}\left\{(r_{\mathrm{r}}^0 - r_{\mathrm{D}})\sigma^{-1}\mathbf{1}\right\}(t) \\
&-\left\{(2r_{\mathrm{D}} - r_{\mathrm{f}}^0 + h_1 + h_2) + (r_{\mathrm{r}}^0 - r_{\mathrm{D}})(\sigma_I + \sigma_C)\sigma^{-1}\mathbf{1}\right\}(t)y \\
&+\left\{h_1 - (r_{\mathrm{f}}^0 - r_{\mathrm{D}}) + (r_{\mathrm{r}}^0 - r_{\mathrm{D}})\sigma_I\sigma^{-1}\mathbf{1}\right\}(t)p_1 \\
&+\left\{h_2 - (r_{\mathrm{f}}^0 - r_{\mathrm{D}}) + (r_{\mathrm{r}}^0 - r_{\mathrm{D}})\sigma_C\sigma^{-1}\mathbf{1}\right\}(t)p_2 \\
&+\alpha\left\{r_{\mathrm{f}}^0(t)\hat{v} \mp r_{\mathrm{col}}^+(t)\hat{v}^{\pm} \pm r_{\mathrm{col}}^-(t)\hat{v}^{\mp}\right\} \\
&\pm\epsilon_{\mathrm{f}}(t)\left|y + (p_1 - y) + (p_2 - y) - \alpha\hat{v}\right| \\
&\pm\epsilon_{\mathrm{r}}(t)\left|\left\{z^{\top} + (p_1 - y)\sigma_I(t) + (p_2 - y)\sigma_C(t)\right\}\sigma(t)^{-1}\mathbf{1}\right|.
\end{aligned}$$

So, for $\delta_0, \delta_1, \delta_2 \geq 0$, we see that

$$\begin{aligned}
&\bar{f}^{+}\left(\cdot, y, z; \hat{v} + \delta_0, p_1 + \delta_1, p_2 + \delta_2\right) - \bar{f}^{+}\left(\cdot, y, z; \hat{v}, p_1, p_2\right) \\
=&\left\{h_1 - (r_{\mathrm{f}}^0 - r_{\mathrm{D}}) + (r_{\mathrm{r}}^0 - r_{\mathrm{D}})\sigma_I\sigma^{-1}\mathbf{1}\right\}\delta_1 \\
&+\left\{h_2 - (r_{\mathrm{f}}^0 - r_{\mathrm{D}}) + (r_{\mathrm{r}}^0 - r_{\mathrm{D}})\sigma_C\sigma^{-1}\mathbf{1}\right\}\delta_2 \\
&+\alpha\left[r_{\mathrm{f}}^0\delta_0 - r_{\mathrm{col}}^+\left\{(\hat{v} + \delta_0)^+ - \hat{v}^+\right\} + r_{\mathrm{col}}^-\left\{(\hat{v} + \delta_0)^- - \hat{v}^-\right\}\right] \\
&+\epsilon_{\mathrm{f}}\left\{\left|p_1 + p_2 - \alpha\hat{v} - y + (\delta_1 + \delta_2 - \alpha\delta_0)\right| - \left|p_1 + p_2 - \alpha\hat{v} - y\right|\right\} \\
&+\epsilon_{\mathrm{r}}\Big[\left|\left\{z^{\top} + (p_1 - y)\sigma_I + (p_2 - y)\sigma_C\right\}\sigma^{-1}\mathbf{1} + \{\delta_1\sigma_I + \delta_2\sigma_C\}\sigma^{-1}\mathbf{1}\right| \\
&-\left|\left\{z^{\top} + (p_1 - y)\sigma_I + (p_2 - y)\sigma_C\right\}\sigma^{-1}\mathbf{1}\right|\Big].
\end{aligned} \tag{50}$$

Using the inequality $|x + y| - |x| \geq -|y|$ and the relation

$$r_{\mathrm{col}}^+\left\{(\hat{v} + \delta_0)^+ - \hat{v}^+\right\} - r_{\mathrm{col}}^-\left\{(\hat{v} + \delta_0)^- - \hat{v}^-\right\} \leq \left(r_{\mathrm{col}}^+ \vee r_{\mathrm{col}}^-\right)\delta_0,$$

we see that

$$\begin{aligned}
(50) \geq&\left\{h_1 - (r_{\mathrm{f}}^0 - r_{\mathrm{D}}) + (r_{\mathrm{r}}^0 - r_{\mathrm{D}})\sigma_I\sigma^{-1}\mathbf{1}\right\}\delta_1 \\
&+\left\{h_2 - (r_{\mathrm{f}}^0 - r_{\mathrm{D}}) + (r_{\mathrm{r}}^0 - r_{\mathrm{D}})\sigma_C\sigma^{-1}\mathbf{1}\right\}\delta_2 + \alpha\left(r_{\mathrm{f}}^0 - r_{\mathrm{col}}^-\right)\delta_0 \\
&-\epsilon_{\mathrm{f}}(\delta_1 + \delta_2 + \alpha\delta_0) - \epsilon_{\mathrm{r}}\left\{|\sigma_I\sigma^{-1}\mathbf{1}|\delta_1 + |\sigma_C\sigma^{-1}\mathbf{1}|\delta_2\right\} \\
=&\left\{h_1 - r_{\mathrm{f}}^- + r_{\mathrm{D}} + (r_{\mathrm{r}}^0 - r_{\mathrm{D}})\sigma_I\sigma^{-1}\mathbf{1} - \epsilon_{\mathrm{r}}|\sigma_I\sigma^{-1}\mathbf{1}|\right\}\delta_1 \\
&+\left\{h_2 - r_{\mathrm{f}}^- + r_{\mathrm{D}} + (r_{\mathrm{r}}^0 - r_{\mathrm{D}})\sigma_C\sigma^{-1}\mathbf{1} - \epsilon_{\mathrm{r}}|\sigma_C\sigma^{-1}\mathbf{1}|\right\}\delta_2 \\
&+\alpha\left(r_{\mathrm{f}}^+ - r_{\mathrm{col}}^-\right)\delta_0 \geq 0,
\end{aligned} \tag{51}$$

where we use (48) and (49). Consider the system of BSDEs (42) and write the solution as

$$\bar{Y}^{\pm}(t; \xi_T, \phi_1, \phi_2), \quad \bar{Z}^{\pm}(t; \xi_T, \phi_1, \phi_2) \quad t \in [0, T]$$

by emphasizing the parameters (ξ_T, ϕ_1, ϕ_2). Take other payoff parameters $\left(\tilde{\xi}_T, \tilde{\phi}_1, \tilde{\phi}_2\right)$ such that $\tilde{\xi}_T \geq \xi_T$, $\tilde{\phi}_1 \geq \phi_1$, and $\tilde{\phi}_2 \geq \phi_2$. Using the comparison theorem for BSDEs twice (for $\hat{V}$ and $\bar{Y}^+$), and using relations (50) and (51), we deduce that

$$\bar{Y}^+\left(\tilde{\xi}_T, \tilde{\phi}_1, \tilde{\phi}_2\right) \geq \bar{Y}^+(\xi_T, \phi_1, \phi_2)$$

and that

$$Y^+\left(\tilde{\xi}_T, \tilde{\phi}_1, \tilde{\phi}_2\right) \geq Y^+(\xi_T, \phi_1, \phi_2).$$

This implies the minimality of $Y^+(\xi_T, \phi_1, \phi_2)$ and the equality,

$$\bar{p} = Y^+(0; \xi_T, \phi_1, \phi_2).$$

The equality,

$$\underline{p} = Y^-(0; \xi_T, \phi_1, \phi_2),$$

can be seen similarly.

Remark 15 We have that for $k \geq 0$,

$$Y^{\pm}(t; k\xi_T, k\phi_1, k\phi_2) \equiv kY^{\pm}(t; \xi_T, \phi_1, \phi_2) \quad \text{for } t \in [0, T].$$

This positive homogeneity is seen from those of the drivers of BSDEs (42), namely

$$\begin{aligned} \bar{f}^{\pm}(t, ky, kz; k\hat{v}, k\phi_1, k\phi_2) &= k\bar{f}^{\pm}(t, y, z; \hat{v}, \phi_1, \phi_2), \\ -r_{\mathrm{D}}(t)(k\hat{v}) &= k\left\{-r_{\mathrm{D}}(t)\hat{v}\right\}. \end{aligned}$$

See Jiang (2008) for the details.

4.2 Results on XVA

In this subsection, we assume that

$$\epsilon_{\mathrm{f}} \vee \epsilon_{\mathrm{r}} \leq \epsilon \tag{52}$$

with some (small) positive constant $\epsilon \ll 1$. Consider the system of BSDEs

$$
\begin{aligned}
-\mathrm{d}Y^{0,\pm}(t) &= f^{0,\pm}\left(t, Y^{0,\pm}(t), Z^{0,\pm}(t), U_1^{0,\pm}(t), U_2^{0,\pm}(t); \hat{V}(t)\right)\mathrm{d}t \\
&\quad - Z^{0,\pm}(t)^\top \mathrm{d}W(t) - U_1^{0,\pm}(t)\mathrm{d}M_1(t) - U_2^{0,\pm}(t)\mathrm{d}M_2(t), \\
&\quad \text{for} \quad 0 \le t \le \tau_1 \wedge \tau_2 \wedge T, \\
Y^{0,\pm}(\tau_1 \wedge \tau_2 \wedge T) &= H, \\
-\mathrm{d}\hat{V}(t) &= -r_{\mathrm{D}}(t)\hat{V}(t)\mathrm{d}t - \Delta(t)^\top \mathrm{d}W(t) \quad \text{for} \quad 0 \le t \le T, \\
\hat{V}(T) &= \xi_T
\end{aligned}
\tag{53}
$$

on $(\Omega, \mathcal{F}, \mathbb{P}, \mathbb{G})$, where

$$
\begin{aligned}
&f^{0,\pm}(t, y, z, u_1, u_2; \hat{v}) \\
&:= f^0(t, y, z, u_1, u_2) + \alpha \left\{ r_{\mathrm{f}}^0(t)\hat{v} \mp r_{\mathrm{col}}^+(t)\hat{v}^\pm \pm r_{\mathrm{col}}^-(t)\hat{v}^\mp \right\}.
\end{aligned}
$$

Associated with (53), consider the reduced system of BSDEs

$$
\begin{aligned}
-\mathrm{d}\bar{Y}^{0,\pm}(t) &= \bar{f}^{0,\pm}\left(t, \bar{Y}^{0,\pm}(t), \bar{Z}^{0,\pm}(t); \hat{V}(t), \phi_1(t), \phi_2(t)\right)\mathrm{d}t \\
&\quad - \bar{Z}^{0,\pm}(t)^\top \mathrm{d}W(t) \quad \text{for} \quad 0 \le t \le T, \\
\bar{Y}^{0,\pm}(T) &= \xi_T, \\
-d\hat{V}(t) &= -r_{\mathrm{D}}(t)\hat{V}(t)\mathrm{d}t - \Delta(t)^\top \mathrm{d}W(t) \quad \text{for} \quad 0 \le t \le T, \\
\hat{V}(T) &= \xi_T
\end{aligned}
\tag{54}
$$

on $(\Omega, \mathcal{F}, \mathbb{P}, \mathbb{F})$, where

$$
\begin{aligned}
&\bar{f}^{0,\pm}(t, y, z; \hat{v}, p_1, p_2) \\
&:= f^{0,\pm}(t, y, z, p_1 - y, p_2 - y; \hat{v}) + (p_1 - y)h_1(t) + (p_2 - y)h_2(t).
\end{aligned}
$$

We obtain the following.

Theorem 5 *Assume Conditions (19) and (44). For $(\bar{Y}^\pm, \bar{Z}^\pm)$, $(\bar{Y}^{0,\pm}, \bar{Z}^{0,\pm})$, which are solutions to BSDEs (42) and (54), respectively, it holds that*

$$
\bar{Y}^- \le \bar{Y}^{0,-} \le \bar{Y}^{0,+} \le \bar{Y}^+ \tag{55}
$$

and that

$$
\left\| \bar{Y}^\pm - \bar{Y}^{0,\pm} \right\|_{\beta,T} + \left\| \bar{Z}^\pm - \bar{Z}^{0,\pm} \right\|_{\beta,T} = O(\epsilon) \tag{56}
$$

as $\epsilon \to 0$ in both $+$ and $-$ cases.

Proof (Sketch). The relation (55) is easily seen from the comparison theorem of BSDEs. To see (56), we can apply the continuity (and the differentiability) results with their proofs with respect to parameterized BSDEs, shown in El Karoui et al. (2000) (see Proposition 2.4 and its proof in El Karoui et al. 2000 for the details).

Combining Theorems 4 and 5, we see the following.

Corollary 1 *Assume Conditions (19), (44), (48), and (49). Then,* $Y^{0,-}(0)$ *and* $Y^{0,+}(0)$ *are arbitrage-free prices at time* 0 *for the derivative security given in Definition 2.*

The above corollary implies that $Y^{0,\pm}(0)$ may be regarded as approximated prices of the derivative security for the writer and her counterparty, which prohibit the existence of an arbitrage opportunity. Because BSDEs for $(Y^{0,\pm}, Z^{0,\pm})$ are linear,[5] we obtain the closed-form expressions for $Y^{0,\pm}$ as follows. Let us introduce the probability measure $\tilde{\mathbb{P}}_T$ on $(\Omega, \mathcal{F}_T)$ by

$$d\tilde{\mathbb{P}}_T \big|_{\mathcal{F}_t} = \mathcal{E}(t) d\mathbb{P} \big|_{\mathcal{F}_t}, \quad t \in [0, T],$$

where

$$\mathcal{E}(t) := \exp\Bigg[\int_0^t \left\{ r_{\mathrm{r}}^0(u) - r_{\mathrm{D}}(u) \right\} \mathbf{1}^\top (\sigma(u)^{-1})^\top \mathrm{d}W(u) - \frac{1}{2} \int_0^t \left\{ r_{\mathrm{r}}^0(u) - r_{\mathrm{D}}(u) \right\}^2 \left| \sigma(u)^{-1} \mathbf{1} \right|^2 du \Bigg].$$

We denote the expectation with respect to $\tilde{\mathbb{P}}_T$ conditioned by $\mathcal{F}_t$ by $\tilde{\mathbb{E}}_t[(\cdots)] = \tilde{\mathbb{E}}[(\cdots)|\mathcal{F}_t]$. Recall that

$$\tilde{W}(t) := W(t) - \int_0^t \left\{ r_{\mathrm{r}}^0(u) - r_{\mathrm{D}}(u) \right\} \sigma(u)^{-1} \mathbf{1} du, \quad t \in [0, T]$$

is a $(\tilde{\mathbb{P}}_T, \mathbb{F})$-Brownian motion by the Maruyama–Girsanov theorem, and on $\left(\Omega, \mathcal{F}, \tilde{\mathbb{P}}_T, \mathbb{F}\right)$ the risky asset price process S has the dynamics

$$dS(t) = \mathrm{diag}(S(t)) \left\{ \sigma(t) d\tilde{W}(t) + r_{\mathrm{r}}^0(t) \mathbf{1} \mathrm{d}t \right\}, \quad S(0) \in \mathbb{R}^n_{++}.$$

Also, we denote

$$\mathrm{DF}_r(t, u) := \exp\left\{ - \int_t^u r(s) \mathrm{d}s \right\}$$

for the process $r := (r(t))_{t \in [0,T]}$. We then obtain the following.

Proposition 3 *The following representation holds:*

[5] That is, the drivers $f^{0,\pm}(t, y, z, u_1, u_2; \hat{v})$ are linear with respect to (y, z, u_1, u_2).

$$\bar{Y}^{0,\pm}(t) = \mathrm{V}(t) + \mathrm{VA}_1(t) + \mathrm{VA}_2(t) + \mathrm{VA}_3(t) + \mathrm{VA}_4(t) + \mathrm{VA}_5^{\pm}(t). \tag{57}$$

Here,

$$\begin{aligned}
\mathrm{V}(t) :=& \tilde{\mathbb{E}}_t\left[\mathrm{DF}_{r_f^0}(t,T)\xi_T\right],\\
\mathrm{VA}_1(t) :=& \tilde{\mathbb{E}}_t\left[\int_t^T \mathrm{DF}_R(t,u)h_1(u)\hat{\phi}_1(u)du\right],\\
\mathrm{VA}_2(t) :=& \tilde{\mathbb{E}}_t\left[\int_t^T \mathrm{DF}_R(t,u)h_2(u)\hat{\phi}_2(u)du\right],\\
\mathrm{VA}_3(t) :=& -\tilde{\mathbb{E}}_t\left[\int_t^T \mathrm{DF}_R(t,u)\left\{(r_f^0 - r_D)\left(\hat{\phi}_1 + \hat{\phi}_2\right)\right\}(u)du\right],\\
\mathrm{VA}_4(t) :=& \tilde{\mathbb{E}}_t\left[\int_t^T \mathrm{DF}_R(t,u)\left\{(r_r^0 - r_D)\left(\hat{\phi}_1\sigma_I + \hat{\phi}_2\sigma_C\right)\sigma^{-1}\mathbf{1}\right\}(u)du\right],\\
\mathrm{VA}_5^{\pm}(t) :=& \alpha\tilde{\mathbb{E}}_t\left[\int_t^T \mathrm{DF}_R(t,u)\left\{\left(r_f^0 - r_{col}^{\pm}\right)\hat{V}^+ - \left(r_f^0 - r_{col}^{\mp}\right)\hat{V}^-\right\}(u)du\right],
\end{aligned}$$

where we define

$$\begin{aligned}
\hat{\phi}_i :=& \phi_i - V \quad \text{for } i = 1, 2, \quad \text{and}\\
R :=& r_D - \left(r_f^0 - r_D\right) + \left\{(r_r^0 - r_D)(\sigma_I + \sigma_C)(\sigma)^{-1}\mathbf{1}\right\} + h_1 + h_2.
\end{aligned}$$

Proof Using the representation formula for linear BSDE (e.g., see Proposition 2.2 of El Karoui et al. 2000), we see that

$$\bar{Y}^{0,\pm}(t) = \bar{\mathrm{V}}(t) + \overline{\mathrm{VA}}_1(t) + \overline{\mathrm{VA}}_2(t) + \overline{\mathrm{VA}}_3(t) + \overline{\mathrm{VA}}_4(t) + \mathrm{VA}_5^{\pm}(t),$$

where

$$\begin{aligned}
\bar{\mathrm{V}}(t) :=& \tilde{\mathbb{E}}_t\left[\mathrm{DF}_R(t,T)\xi_T\right],\\
\overline{\mathrm{VA}}_1(t) :=& \tilde{\mathbb{E}}_t\left[\int_t^T \mathrm{DF}_R(t,u)h_1(u)\phi_1(u)\mathrm{d}u\right],\\
\overline{\mathrm{VA}}_2(t) :=& \tilde{\mathbb{E}}_t\left[\int_t^T \mathrm{DF}_R(t,u)h_2(u)\phi_2(u)\mathrm{d}u\right],
\end{aligned}$$

$$\overline{\mathrm{VA}}_3(t) := -\tilde{\mathbb{E}}_t\left[\int_t^T \mathrm{DF}_R(t,u)\left\{(r_{\mathrm{f}}^0 - r_{\mathrm{D}})(\phi_1+\phi_2)\right\}(u)\mathrm{d}u\right],$$

$$\overline{\mathrm{VA}}_4(t) := \tilde{\mathbb{E}}_t\left[\int_t^T \mathrm{DF}_R(t,u)\left\{(r_{\mathrm{r}}^0 - r_{\mathrm{D}})(\phi_1\sigma_I+\phi_2\sigma_C)\sigma^{-1}\mathbf{1}\right\}(u)\mathrm{d}u\right].$$

Furthermore, we see that

$$\begin{aligned}
&\left[\mathrm{VA}_1+\mathrm{VA}_2+\mathrm{VA}_3+\mathrm{VA}_4-\overline{\mathrm{VA}}_1-\overline{\mathrm{VA}}_2-\overline{\mathrm{VA}}_3-\overline{\mathrm{VA}}_4\right](t)\\
&=-\tilde{\mathbb{E}}_t\left[\int_t^T \mathrm{DF}_R(t,u)\mathrm{V}(u)\left\{R(u)-r_{\mathrm{f}}^0(u)\right\}\mathrm{d}u\right]\\
&=-\tilde{\mathbb{E}}_t\left[\int_t^T \mathrm{DF}_R(t,u)\tilde{\mathbb{E}}_u\left[\mathrm{DF}_{r_{\mathrm{f}}^0}(u,T)\xi_T\right]\left\{R(u)-r_{\mathrm{f}}^0(u)\right\}\mathrm{d}u\right]\\
&=\tilde{\mathbb{E}}_t\left[\mathrm{DF}_{r_{\mathrm{f}}^0}(t,T)\xi_T\int_t^T \frac{\partial}{\partial u}\mathrm{DF}_{R-r_{\mathrm{f}}^0}(t,u)\mathrm{d}u\right]\\
&=\tilde{\mathbb{E}}_t\left[\mathrm{DF}_{r_{\mathrm{f}}^0}(t,T)\left\{\mathrm{DF}_{R-r_{\mathrm{f}}^0}(t,T)-1\right\}\xi_T\right]\\
&=\tilde{\mathbb{E}}_t\left[\left\{\mathrm{DF}_R(t,T)-\mathrm{DF}_{r_{\mathrm{f}}^0}(t,T)\right\}\xi_T\right]=\bar{\mathrm{V}}(t)-\mathrm{V}(t),
\end{aligned}$$

hence the proof is complete.

Remark 16 Suppose that $r_{\mathrm{r}}^0 \equiv r_{\mathrm{f}}^0 \equiv r_{\mathrm{D}}$ holds. In this case, $\tilde{\mathbb{P}}_T \equiv \mathbb{P}$ and $\mathrm{V} \equiv \hat{V}$ follow. Furthermore, consider $\phi_i(t) := \varphi_i\left(\hat{V}(t)\right)$, where (17) is employed for $i = 1, 2$. Then, in (57), $\mathrm{VA}_3 \equiv \mathrm{VA}_4 \equiv 0$, and $-\mathrm{VA}_1$, VA_2, and $\mathrm{VA}_5^{\pm}$ are called the debt valuation adjustment (DVA), the credit valuation adjustment (CVA), and the collateral valuation adjustment (ColVA), respectively, which are popularly used XVA terms in practice for the valuation adjustment in the pricing of derivative securities. Concretely, DVA, CVA, and ColVA at time t are written as

$$\mathrm{DVA}(t) := -\mathbb{E}_t\left[\int_t^T \mathrm{DF}_{r_{\mathrm{D}}+h_1+h_2}(t,u)h_1(u)\hat{\phi}_1(u)\mathrm{d}u\right],$$

$$\mathrm{CVA}(t) := \mathbb{E}_t\left[\int_t^T \mathrm{DF}_{r_{\mathrm{D}}+h_1+h_2}(t,u)h_2(u)\hat{\phi}_2(u)\mathrm{d}u\right],$$

$$\mathrm{ColVA}^{\pm}(t) :=$$

$$\mathbb{E}_t\left[\int_t^T \mathrm{DF}_{r_{\mathrm{D}}+h_1+h_2}(t,u)\left\{\left(r_{\mathrm{D}}-r_{\mathrm{col}}^{\pm}\right)\alpha\hat{V}^{+}-\left(r_{\mathrm{D}}-r_{\mathrm{col}}^{\mp}\right)\alpha\hat{V}^{-}\right\}(u)\mathrm{d}u\right],$$

respectively, where we denote $\mathbb{E}_t[(\cdots)] := \mathbb{E}[(\cdots)|\mathcal{F}_t]$. Further,

$$\mathrm{FVA}(t) := \mathbb{E}_t\left[\int_t^T \mathrm{DF}_{r_{\mathrm{D}}+h_1+h_2}(t,u)\left\{(r_{\mathrm{f}}^0-r_{\mathrm{D}})(\phi_1+\phi_2)\right\}(u)\mathrm{d}u\right],$$

called the funding valuation adjustment (FVA) at time t, is another popularly used adjustment term in practice, which reflects the funding cost of uncollateralised derivatives above the risk-free rate of return. We can roughly relate these XVA terms with the correction terms in Proposition 3 as follows: Let $r_{\mathrm{r}}^0 \equiv r_{\mathrm{D}}$,[6] which implies $\mathrm{VA}_4 \equiv 0$. Further, suppose $r_{\mathrm{f}}^0 \approx r_{\mathrm{D}}$. Then, we may interpret as

$$\begin{aligned}\mathrm{DVA} &\approx -\,\mathrm{VA}_1,\\ \mathrm{CVA} &\approx \mathrm{VA}_2,\\ \mathrm{ColVA}^{\pm} &\approx \mathrm{VA}_5^{\pm},\end{aligned}$$

and

$$\mathrm{FVA} \approx \mathrm{VA}_3,$$

or

$$\mathrm{FVA} \approx \mathrm{VA}_3 + (\mathrm{VA}_1 + \mathrm{DVA}) + (\mathrm{VA}_2 - \mathrm{CVA}) + \left(\mathrm{ColVA}^{\pm} - \mathrm{VA}_5^{\pm}\right).$$

For other theoretical studies on the valuation adjustments and related interpretation of XVA used in practice, we refer to Brigo et al. (2020) and the reference therein. Also, for comprehensive information on XVA issue and expanding-related issues (e.g., computational issue), see for example Gregory (2015) and Glau et al. (2016), and the references therein, which are still nonexhaustive.

4.3 Perturbed BSDEs

As we see in Theorem 5 and Corollary 1, under certain conditions, $Y^{0,+}(t)(< Y^{+}(t))$, which is a zeroth-order approximation of the minimal hedging cost $Y^{+}(t)$, is an arbitrage-free price for the writer at time t. In this subsection, we try to improve our hedging strategy by using a first-order approximation. Using the solution to BSDE (53), consider the linear BSDE

[6] In practice, the difference $r_{\mathrm{r}}^0 - r_{\mathrm{D}}$ seems to have been usually ignored.

$$
\begin{aligned}
-\mathrm{d}Y^{1,\pm}(t) &= f^0\left(t, Y^{1,\pm}(t), Z^{1,\pm}(t), U_1^{1,\pm}(t), U_2^{1,\pm}(t)\right)\mathrm{d}t \\
&\quad + f^{1,\pm}\left(t, Y^{0,\pm}(t), Z^{0,\pm}(t), U_1^{0,\pm}(t), U_2^{0,\pm}(t), \hat{V}(t)\right)\mathrm{d}t \\
&\quad - Z^{1,\pm}(t)\mathrm{d}W(t) - U_1^{1,\pm}(t)\mathrm{d}M_1(t) - U_2^{1,\pm}(t)\mathrm{d}M_2(t), \\
Y^{1,\pm}(\tau_1 \wedge \tau_2 \wedge T) &= 0
\end{aligned}
\tag{58}
$$

on $(\Omega, \mathcal{F}, \mathbb{P}, \mathbb{G})$, where

$$
\begin{aligned}
f^{1,\pm}(t, y, z, u_1, u_2; \hat{v}) &:= \pm\epsilon_{\mathrm{f}}(t)\left|y + u_1 + u_2 - \alpha\hat{v}\right| \\
&\pm\epsilon_{\mathrm{r}}(t)\left|\left\{z^\top + u_1\sigma_I(t) + u_2\sigma_C(t)\right\}\sigma(t)^{-1}\mathbf{1}\right|.
\end{aligned}
$$

Furthermore, using the solution to BSDE (54), consider the linear BSDE

$$
\begin{aligned}
-\mathrm{d}\bar{Y}^{1,\pm}(t) &= \bar{f}^0\left(t, \bar{Y}^{1,\pm}(t), \bar{Z}^{1,\pm}(t); \phi_1(t), \phi_2(t)\right)\mathrm{d}t \\
&\quad + \bar{f}^{1,\pm}\left(t, \bar{Y}^{0,\pm}(t), \bar{Z}^{0,\pm}(t); \hat{V}(t), \phi_1(t), \phi_2(t)\right)\mathrm{d}t \\
&\quad - \bar{Z}^{1,\pm}(t)\mathrm{d}W(t), \\
\bar{Y}^{1,\pm}(T) &= 0
\end{aligned}
\tag{59}
$$

on $(\Omega, \mathcal{F}, \mathbb{P}, \mathbb{F})$, where

$$
\begin{aligned}
\bar{f}^0(t, y, z; p_1, p_2) &:= f^0\left(t, y, z, p_1 - y, p_2 - y\right), \\
\bar{f}^{1,\pm}(t, y, z; \hat{v}, p_1, p_2) &:= \pm\,\epsilon_{\mathrm{f}}(t)\left|y + (p_1 - y) + (p_2 - y) - \alpha\hat{v}\right| \\
&\quad \pm\,\epsilon_{\mathrm{r}}(t)\left|\left\{z^\top + (p_1 - y)\sigma_I(t) + (p_2 - y)\sigma_C(t)\right\}\sigma(t)^{-1}\mathbf{1}\right|.
\end{aligned}
$$

Using a similar technique to that used in the proof of Theorem 5, we can show the following.

Proposition 4 *It holds that for any sufficiently large $\beta > 0$,*

$$
\|\bar{Y}^\pm - (\bar{Y}^{0,\pm} + \bar{Y}^{1,\pm})\|_{\beta,T} + \|\bar{Z}^\pm - (\bar{Z}^{0,\pm} + \bar{Z}^{1,\pm})\|_{\beta,T} = O(\epsilon^2)
$$

as $\epsilon \to 0$, where we assume (52).

Acknowledgements The authors are grateful to an anonymous referee for valuable comments and helpful suggestions.

References

Aksamit A, Jeanblanc M (2017) Enlargement of filtration with finance in view. Springer briefs in quantitative finance. Springer

Bichuch M, Capponi A, Sturm S (2015) Arbitrage-free pricing of XVA–part II: PDE representations and numerical analysis. In: Working paper. Available at http://ssrn.com/abstract=2568118

Bichuch M, Capponi A, Sturm S (2018) Arbitrage-free XVA. Math Finance 28:582–620

Bielecki TR, Cialenco I, Rutkowski M (2018) Arbitrage-free pricing of derivatives in nonlinear market models. Prob Uncertain Quant Risk 3(2):1–56. https://doi.org/10.1186/s41546-018-0027-x

Bielecki TR, Jeanblanc M, Rutkowski M (2005) PDE approach to valuation and hedging of credit derivatives. Quant Finance 5(3):257–270

Bielecki TR, Rutkowski M (2004) Credit risk: modeling, valuation, and hedging. Springer (2004)

Bismut JM (1976) Linear quadratic optimal stochastic control with random coefficients. SIAM J Control Opt 14(3):419–444

Bismut JM (1978) An introductory approach to duality in optimal stochastic control. SIAM Rev 20:62–78

Brigo D, Buescu C, Francischello M, Pallavicini A, Rutkowski M (2020) Nonlinear valuation with XVAs: two converging approaches. Preprint

Cohen SN Elliott RJ (2015) Stochastic calculus and applications, 2nd edn. Birkhäuser

Crépey S (2015) Bilateral counterparty risk under funding constraints-part II: CVA. Math Finance 25(1):23–50

Crépey S, Song S (2016) Counterparty risk and funding: immersion and beyond. Finance Stoch 20:901–930

Darling RWR, Pardoux E (1997) Backward SDE with random terminal time and applications to semilinear elliptic PDE. Ann Prob 25(3):1135–1159

El Karoui N, Huang S-J (1997) A general result of existence and uniqueness of backward stochastic differential equations. In: Backward stochastic differential equations (Pitman research notes in mathematics series, vol 364), pp 27–36

El Karoui N, Peng S, Quenez MC (2000) Backward stochastic differential equations in finance. Math Finance 7(1):1–71

Fujii M, Takahashi A (2011) Choice of collateral currency. Risk 24(1):120–125

Glau K, Grbac Z, Scherer M, Zagst R (2016) Innovations in derivatives markets. In: Springer proceedings in mathematics & statistics, vol 165, Springer

Gregory J (2015) The XVA Challenge: counterparty credit risk, funding, collateral and capital. Wiley Finance (2015)

Jiang L (2008) Convexity, translation invariance and subadditivity for g-expectations and related risk measures. Ann Appl Prob 18(1):245–258

Nagayama Y (2019) Jump-type backward stochastic differential equations with stochastic Lipschitz coefficient and their applications. Master thesis, Graduate School of Engineering Science, Osaka University (in Japanese)

Nie T, Rutkowski M (2018) Fair bilateral pricing under funding costs and exogenous collateralization. Math Finance 28:621–655

Pardoux E, Peng S (1990) Adapted solution of backward stochastic equation. Syst Control Lett 14:55–61

Pham H (2010) Stochastic control under progressive enlargement of filtrations and applications to multiple defaults risk management. Stoch Process Appl 120:1795–1820

Tanaka A (2019) Remarks on an arbitrage-free condition for XVA. JSIAM Lett 11:57–60

Thoednithi K (2015) Some results from arbitrage opportunity on nonlinear wealth processes. J Trans Inst Syst Control Inf Eng 28(7):291–298

Zhang J (2017) Backward stochastic differential equations: from linear to fully nonlinear theory. Springer

An Overview of Exact Solution Methods for Guaranteed Minimum Death Benefit Options in Variable Annuities

Eric R. Ulm

1 Introduction

In his book, "The Calculus of Retirement Income", Milevsky (2006) describes the guaranteed minimum death benefit (GMDB) option in variable annuity products. He describes a gap in the literature with the words, "it is very difficult to obtain a closed-form solution" for GMDB options and to date not many exist (pg 259). Since that time, a number of closed-form solutions have been obtained for some specific option features and some fairly general mortality laws. This paper fills the gap identified by Milevsky by compiling the methods and solutions that have accumulated in the literature since this statement was made in 2006.

In its most basic form, a GMDB is a European option with a random exercise time drawn from the probability density function of an individual's remaining lifetime. The underlying is a fund invested in a risky asset. The owner invests a premium X into the fund. The fund value at any time t is given as S_t which is typically assumed to follow a geometric Brownian motion process. The individual can surrender the fund at any time for the value S_t. If the individual dies, however, the beneficiary receives the maximum of S_t and the strike X_t which is some function of the original premium and the past behavior of the fund. This is equivalent to the beneficiary receiving the fund as well as a European option exercisable at the moment of death.

The simplest GMDB to solve is the "Return of Premium" GMDB. In this case, the strike $X_t = X$, the original strike. The GMDB is then equivalent to a European put option with a random exercise time. A "Roll-up" GMDB allows the strike to increase with time as $X_t = Xe^{\rho t}$. The GMDB is then equivalent to a European put with a strike level dependent on the random exercise time. Finally, a "Ratchet" GMDB allows the strike to move upward with time as the fund increases, but not downward. In other words, the strike at the random exercise time is the maximum of the historical fund

E. R. Ulm (✉)
Victoria University of Wellington, Wellington 6011, New Zealand
e-mail: Eric.Ulm@vuw.ac.nz

J. Cheng et al. (eds.), *Proceedings of the Forum "Math-for-Industry" 2018*,
Mathematics for Industry 35, https://doi.org/10.1007/978-981-16-5576-0_3

value. In the case of a continuous ratchet, the strike is the historical maximum and in the case of a discrete ratchet, it is the historical maximum at all ratchet dates. The GMDB is then equivalent to a lookback option with a random exercise time.

In addition to the features mentioned above, the owner of the fund has several "real options". First, the owner has the right to ask for part or all of the fund value at any time, called a "surrender" or "lapse". Second, there are often a number of funds underlying the option and the owner can choose his allocation and thereby affect the fund's expected return and volatility. If these options were traded in a complete market, one could assume the real options are exercised optimally in order to maximum the risk-neutral value of the option. However, the options are not tradeable in markets, hedging is expensive for ordinary policyholders, and individuals are not able to fully diversify their own mortality risk. For these reasons, it is often assumed that individuals make these choices to maximize the present value of their own and their beneficiary utility (see, for example, Gao and Ulm (2012), Moenig (2012), Gao and Ulm (2015), and Moenig and Zhu (2018)).

Three main methods have been used to obtain exact solutions to these option values. First, the value of the option can be determined by taking the integral of the European option value multiplied by the pdf of the owner's remaining lifetime. Second, the partial differential equation satisfied by the option value can be solved. Third, the option price can be determined using expectations under the risk-neutral probability measure using a discounted density approach. These three methods will be discussed in turn in the following sections.

2 The Direct Integration Method

Direct integration was the first method used to obtain exact solutions for GMDB options. Milevsky and Posner (2001) were the first to solve a specific case of the GMDB option values. They were interested in the at-the-money values. This is not as restrictive as it might seem at first, since nearly all GMDB options are issued at-the-money and the initial option prices can be determined from the at-the-money option values. These options are typically paid for by deduction of fees as a percentage of fund.

In particular, they analyze return-of-premium, roll-up, and ratchet GMDB options with a constant force of mortality (i.e., an exponential distribution for the future lifetime). They evaluate the integral of the European put prices multiplied by the pdf of the future lifetime. Assuming a constant morality rate μ, the value of the roll-up GMDB option can be expressed as:

$$f = \int_0^T X\mathrm{e}^{(p-r)t} N\left(-\xi_2\sqrt{t}\right)\mu\mathrm{e}^{-\mu t}\mathrm{d}t - \int_0^T X\mathrm{e}^{-qt} N\left(-\xi_1\sqrt{t}\right)\mu\mathrm{e}^{-\mu t}\mathrm{d}t \quad (1)$$

where p is the roll-up rate, r is the risk-free rate, q is the rate at which fees are deducted, and T is the maximum time that the option is in effect and

$$\xi_1 = \frac{r - p - q + \frac{\sigma^2}{2}}{\sigma} \text{ and } \xi_2 = \frac{r - p - q - \frac{\sigma^2}{2}}{\sigma} \tag{2}$$

The authors replace the cumulative normal functions with their integral definitions, then reverse the order of integration and take the limit as $T \to \infty$ to find

$$f = \frac{\mu X}{2(r - p + \mu)}\left[1 - \frac{\xi_2}{\sqrt{\xi_2^2 + 2(r - p - \mu)}}\right] - \frac{\mu X}{2(q + \mu)}\left[1 - \frac{\xi_1}{\sqrt{\xi_1^2 + 2(q - \mu)}}\right] \tag{3}$$

This reduces to the return-of-premium formula if $p = 0$. The authors do not give explicit expressions for finite values of T, but they are easy to derive from the other formulas in the paper.

Similarly, the value of a ratchet GMDB option can be expressed as an integral over the known value of the lookback put option found in Goldman et al. (1979). The resulting integral is:

$$\begin{aligned} f = &\int_0^T X e^{-rt} N\left(-\xi_2\sqrt{t}\right)\mu e^{-\mu t}\,dt - \int_0^T X e^{-qt} N\left(-\xi_1\sqrt{t}\right)\mu e^{-\mu t}\,dt \\ &+ \int_0^T \eta X e^{-qt} N\left(\xi_1\sqrt{t}\right)\mu e^{-\mu t}\,dt - \int_0^T \eta X e^{-rt} N\left(\xi_3\sqrt{t}\right)\mu e^{-\mu t}\,dt \end{aligned} \tag{4}$$

where

$$\xi_1 = \frac{r - q + \frac{\sigma^2}{2}}{\sigma};\ \xi_2 = \frac{r - q - \frac{\sigma^2}{2}}{\sigma};\ \xi_3 = \frac{-(r - q) + \frac{\sigma^2}{2}}{\sigma};\ \eta = \frac{\sigma^2}{2(r - q)} \tag{5}$$

Using the same technique of reversing integration order and letting $T \to \infty$, they find:

$$f = \frac{\mu(1 - \eta)X}{2(r + \mu)}\left[1 - \frac{\xi_2}{\sqrt{\xi_2^2 + 2(r + \mu)}}\right] - \frac{\mu X}{2(q + \mu)}\left[1 - \eta - \frac{\xi_1(1 + \eta)}{\sqrt{\xi_1^2 + 2(q + \mu)}}\right] \tag{6}$$

This technique could presumably be applied to any GMDB option whose analogous European option formula contains forms of $N(c\sqrt{t})$. A sampling of such options can be found in the book by Haug (2007). However, most of these solutions would be of theoretical interest only since no such GMDB options are offered in the market. To the best of my knowledge, this methodology has not been employed elsewhere.

3 The Partial Differential Equation Method

It can be shown that the value of the GMDB option must satisfy a partial differential equation in time and asset level. The equation that must be satisfied by the value of the GMDB option, $f_a(S,t)$, if $S \leq X$, is:

$$\frac{\partial f_a}{\partial t} + (r-q)S\frac{\partial f_a}{\partial S} + \frac{1}{2}\sigma^2 S^2 \frac{\partial^2 f_a}{\partial S^2} = [r + \mu_x(t) + \lambda(S,t)]f_a - [\mu_x(t)]\text{Max}(X_t - S, 0) \tag{7}$$

where $\lambda(S,t)$ is the (possibly time and level dependent) lapse rate.

The equation can be derived in a number of ways. Milevsky and Salisbury (2001) derive the equation using standard techniques obtained from the generator of the diffusion process. It can also be shown that the integral formulation satisfies Eq. (7) if the individual put options in the integral obey the standard Black–Scholes PDE. Intuitively, if the market is complete, the value of the option must grow at the risk-free rate in expectation. This implies that the value of the option when the owner is alive must grow faster than the risk-free rate to compensate for its reduced value when the owner is dead producing the first term on the RHS of Eq. (7). The second term reflects a reduction in the growth of the alive option when there are death benefits involved.

If the lapse rate and force of mortality are assumed to be constant, the PDE reduces to an ODE:

$$(r-q)S\frac{\partial f_a}{\partial S} + \frac{1}{2}\sigma^2 S^2 \frac{\partial^2 f_a}{\partial S^2} = [r + \mu + \lambda]f_a - \mu\text{Max}(X_t - S, 0) \tag{8}$$

This equation can be solved straightforwardly in the case of a return of premium GMDB with appropriate boundary conditions. The solution is given in Ulm (2006) as:

$$f_a(S,t) = \left\{ \frac{\mu X}{2(r+\mu+\lambda)}\left[1 - \frac{\xi_2}{\sqrt{\xi_2^2 + 2(r+\mu+\lambda)}}\right] - \frac{\mu X}{2(q+\mu+\lambda)}\left[1 - \frac{\xi_1}{\sqrt{\xi_1^2 + 2(q+\mu+\lambda)}}\right]\right\}\left(\frac{S}{X}\right)^{m_2} \quad S > X \tag{9}$$

and

$$f_a(S,t) = \frac{\mu X}{r+\mu+\lambda} - \frac{\mu X}{q+\mu+\lambda}\left(\frac{S}{X}\right)$$

$$+\left\{\frac{\mu X}{2(q+\mu+\lambda)}\left[1+\frac{\xi_1}{\sqrt{\xi_1^2+2(q+\mu+\lambda)}}\right]\right.$$

$$\left.-\frac{\mu X}{2(r+\mu+\lambda)}\left[1+\frac{\xi_2}{\sqrt{\xi_2^2+2(r+\mu+\lambda)}}\right]\right\}\left(\frac{S}{X}\right)^{m_1} \quad S < X \quad (10)$$

where

$$\xi_1 = \frac{r-q+\frac{\sigma^2}{2}}{\sigma};\ \xi_2 = \frac{r-q-\frac{\sigma^2}{2}}{\sigma};$$

$$m_1 = \frac{-\left(r-q-\frac{\sigma^2}{2}\right)+\sqrt{\left(r-q-\frac{\sigma^2}{2}\right)^2+2\sigma^2(r+\mu+\lambda)}}{\sigma^2};$$

$$m_2 = \frac{-\left(r-q-\frac{\sigma^2}{2}\right)-\sqrt{\left(r-q-\frac{\sigma^2}{2}\right)^2+2\sigma^2(r+\mu+\lambda)}}{\sigma^2} \quad (11)$$

This clearly reduces to Eq. (3) when $\lambda = 0$ and $S = X$.

This equation can be used to determine optimal surrender and allocation strategies. Milevsky and Salisbury (2001) solve the PDE in the presence of surrender charges and find an optimal exercise boundary above which the owner will surrender the policy and reinstate it at another company as a new at-the-money option. They then compute the surrender charges necessary to prevent optimal lapsation from occurring.

Ulm (2006) solves the PDE in the case where the owner has a risky and risk-free asset and finds an optimal exercise boundary. Below the boundary, the fund will be invested entirely in the risk-free account and above the boundary the fund will be invested entirely in the risky asset.

The analytic solution of the PDE for a roll-up GMDB option under exponential mortality has been found in Ulm (2008). The solution is quite similar to that in Eqs. (9)–(11) with some simple redefinitions and is given as:

$$f_a(S,t) = \left\{\frac{\mu X e^{pt}}{2(r+\mu+\lambda-p)}\left[1-\frac{\xi_2}{\sqrt{\xi_2^2+2(r+\mu+\lambda-p)}}\right]\right.$$

$$\left.-\frac{\mu X e^{pt}}{2(q+\mu+\lambda)}\left[1-\frac{\xi_1}{\sqrt{\xi_1^2+2(q+\mu+\lambda)}}\right]\right\}\left(\frac{S}{Xe^{pt}}\right)^{m_2} \quad S > Xe^{pt}$$

(12)

and

$$f_a(S,t) = \frac{\mu X e^{pt}}{r+\mu+\lambda-p} - \frac{\mu X e^{pt}}{q+\mu+\lambda}\left(\frac{S}{Xe^{pt}}\right)$$

$$+\left\{\frac{\mu X e^{pt}}{2(q+\mu+\lambda)}\left[1+\frac{\xi_1}{\sqrt{\xi_1^2+2(q+\mu+\lambda)}}\right]\right.$$

$$\left.-\frac{\mu X e^{pt}}{2(r+\mu+\lambda-p)}\left[1+\frac{\xi_2}{\sqrt{\xi_2^2+2(r+\mu+\lambda-p)}}\right]\right\}\left(\frac{S}{Xe^{pt}}\right)^{m_1} \quad S < Xe^{pt} \tag{13}$$

where

$$\xi_1 = \frac{r-q-p+\frac{\sigma^2}{2}}{\sigma};\ \xi_1 = \frac{r-q-p-\frac{\sigma^2}{2}}{\sigma};$$

$$m_1 = \frac{-\left(r-q-p-\frac{\sigma^2}{2}\right)+\sqrt{\left(r-q-p-\frac{\sigma^2}{2}\right)^2+2\sigma^2(r+\mu+\lambda-p)}}{\sigma^2};$$

$$m_2 = \frac{-\left(r-q-p-\frac{\sigma^2}{2}\right)-\sqrt{\left(r-q-p-\frac{\sigma^2}{2}\right)^2+2\sigma^2(r+\mu+\lambda-p)}}{\sigma^2} \tag{14}$$

Finding analytic solutions only in the case of a constant force of mortality is not as restrictive as it might seem. It can be shown that any distribution can be approximated by a sum of exponentials (see Dufresne (2007)). Since the PDE (Eq. (7)) is linear, the solution for an arbitrary distribution would be the sum of the solutions for the approximating exponentials. Unfortunately, this methodology seems to work poorly in practice, requiring a large number of terms for convergence.

Ulm (2008) also finds solutions for three additional mortality laws where $\mu_x(t)$ is time dependent. First, a solution is obtained for the situation where μ is constant until time T, at which point everyone dies. The solution is given by:

$$f_a(S,t) = \frac{(r+\lambda-p)Xe^{pt}}{r+\mu+\lambda-p} e^{-(r+\mu+\lambda-p)(T-t)} N(-d_2)$$

$$-\frac{(q+\lambda)Xe^{pt}}{q+\mu+\lambda}\left(\frac{S}{Xe^{pt}}\right) e^{-(q+\mu+\lambda)(T-t)} N(-d_1)$$

$$-\left\{\frac{\mu X e^{pt}}{2(q+\mu+\lambda)}\left[1+\frac{\xi_1}{\sqrt{\xi_1^2+2(q+\mu+\lambda)}}\right]\right.$$

$$\left.-\frac{\mu X e^{pt}}{2(r+\mu+\lambda-p)}\left[1+\frac{\xi_2}{\sqrt{\xi_2^2+2(r+\mu+\lambda-p)}}\right]\right\}\left(\frac{S}{Xe^{pt}}\right)^{m_1} N(-d_3)$$

$$+\left\{\frac{\mu X e^{pt}}{2(r+\mu+\lambda-p)}\left[1-\frac{\xi_2}{\sqrt{\xi_2^2+2(r+\mu+\lambda-p)}}\right]\right.$$

$$\left.-\frac{\mu X e^{pt}}{2(q+\mu+\lambda)}\left[1-\frac{\xi_1}{\sqrt{\xi_1^2+2(q+\mu+\lambda)}}\right]\right\}\left(\frac{S}{Xe^{pt}}\right)^{m_2} N(-d_4) \quad S > Xe^{pt} \tag{15}$$

and

$$
\begin{aligned}
f_a(S,t) = {} & \frac{\mu X e^{pt}}{r+\mu+\lambda-p} - \frac{\mu X e^{pt}}{q+\mu+\lambda}\left(\frac{S}{Xe^{pt}}\right) \\
& + \frac{(r+\lambda-p)Xe^{pt}}{r+\mu+\lambda-p} e^{-(r+\mu+\lambda-p)(T-t)} N(-d_2) \\
& - \frac{(q+\lambda)Xe^{pt}}{q+\mu+\lambda}\left(\frac{S}{Xe^{pt}}\right) e^{-(q+\mu+\lambda)(T-t)} N(-d_1) \\
& + \left\{\frac{\mu X e^{pt}}{2(q+\mu+\lambda)}\left[1+\frac{\xi_1}{\sqrt{\xi_1^2+2(q+\mu+\lambda)}}\right]\right. \\
& \left. - \frac{\mu X e^{pt}}{2(r+\mu+\lambda-p)}\left[1+\frac{\xi_2}{\sqrt{\xi_2^2+2(r+\mu+\lambda-p)}}\right]\right\}\left(\frac{S}{Xe^{pt}}\right)^{m_1} N(d_3) \\
& - \left\{\frac{\mu X e^{pt}}{2(r+\mu+\lambda-p)}\left[1-\frac{\xi_2}{\sqrt{\xi_2^2+2(r+\mu+\lambda-p)}}\right]\right. \\
& \left. - \frac{\mu X e^{pt}}{2(q+\mu+\lambda)}\left[1-\frac{\xi_1}{\sqrt{\xi_1^2+2(q+\mu+\lambda)}}\right]\right\}\left(\frac{S}{Xe^{pt}}\right)^{m_2} N(d_4) \quad S < Xe^{pt}
\end{aligned} \tag{16}
$$

with additional definitions:

$$
\begin{aligned}
d_1 &= \frac{\ln\left(\frac{S}{Xe^{pt}}\right) + \left(r-q-p+\frac{\sigma^2}{2}\right)(T-t)}{\sigma\sqrt{T-t}}; \\
d_2 &= \frac{\ln\left(\frac{S}{Xe^{pt}}\right) + \left(r-q-p-\frac{\sigma^2}{2}\right)(T-t)}{\sigma\sqrt{T-t}}; \\
d_3 &= \frac{\ln\left(\frac{S}{Xe^{pt}}\right) + \sigma\sqrt{\xi_1^2+2(q+\mu+\lambda)}(T-t)}{\sigma\sqrt{T-t}}; \\
d_4 &= \frac{\ln\left(\frac{S}{Xe^{pt}}\right) - \sigma\sqrt{\xi_1^2+2(q+\mu+\lambda)}(T-t)}{\sigma\sqrt{T-t}}
\end{aligned} \tag{17}
$$

Second, a solution is obtained for the situation where μ is constant until time T, at which point the option expires worthless. The solution is given by:

$$
\begin{aligned}
f_a(S,t) = {} & \frac{\mu X e^{pt}}{q+\mu+\lambda}\left(\frac{S}{Xe^{pt}}\right) e^{-(q+\mu+\lambda)(T-t)} N(-d_1) \\
& - \frac{\mu X e^{pt}}{r+\mu+\lambda-p} e^{-(r+\mu+\lambda-p)(T-t)} N(-d_2) \\
& - \left\{\frac{\mu X e^{pt}}{2(q+\mu+\lambda)}\left[1+\frac{\xi_1}{\sqrt{\xi_1^2+2(q+\mu+\lambda)}}\right]\right. \\
& \left. - \frac{\mu X e^{pt}}{2(r+\mu+\lambda-p)}\left[1+\frac{\xi_2}{\sqrt{\xi_2^2+2(r+\mu+\lambda)}}\right]\right\}\left(\frac{S}{Xe^{pt}}\right)^{m_1} N(-d_3)
\end{aligned}
$$

$$+\left\{\frac{\mu X e^{pt}}{2(r+\mu+\lambda-p)}\left[1-\frac{\xi_2}{\sqrt{\xi_2^2+2(r+\mu+\lambda-p)}}\right]\right.$$
$$\left.-\frac{\mu X e^{pt}}{2(q+\mu+\lambda)}\left[1-\frac{\xi_1}{\sqrt{\xi_1^2+2(q+\mu+\lambda)}}\right]\right\}\left(\frac{S}{X e^{pt}}\right)^{m_2} N(-d_4) \quad S > X e^{pt} \tag{18}$$

and

$$\begin{aligned} f_a(S,t) = {} & \frac{\mu X e^{pt}}{r+\mu+\lambda-p} - \frac{\mu X e^{pt}}{q+\mu+\lambda}\left(\frac{S}{X e^{pt}}\right) \\ & + \frac{\mu X e^{pt}}{q+\mu+\lambda}\left(\frac{S}{X e^{pt}}\right) e^{-(q+\mu+\lambda)(T-t)} N(-d_1) \\ & - \frac{\mu X e^{pt}}{r+\mu+\lambda-p} e^{-(r+\mu+\lambda-p)(T-t)} N(-d_2) \\ & + \left\{\frac{\mu X e^{pt}}{2(q+\mu+\lambda)}\left[1+\frac{\xi_1}{\sqrt{\xi_1^2+2(q+\mu+\lambda)}}\right]\right. \\ & \left. - \frac{\mu X e^{pt}}{2(r+\mu+\lambda-p)}\left[1+\frac{\xi_2}{\sqrt{\xi_2^2+2(r+\mu+\lambda-p)}}\right]\right\}\left(\frac{S}{X e^{pt}}\right)^{m_1} N(d_3) \\ & - \left\{\frac{\mu X e^{pt}}{2(r+\mu+\lambda-p)}\left[1-\frac{\xi_2}{\sqrt{\xi_2^2+2(r+\mu+\lambda-p)}}\right]\right. \\ & \left. - \frac{\mu X e^{pt}}{2(q+\mu+\lambda)}\left[1-\frac{\xi_1}{\sqrt{\xi_1^2+2(q+\mu+\lambda)}}\right]\right\}\left(\frac{S}{X e^{pt}}\right)^{m_2} N(d_4) \quad S < X e^{pt} \end{aligned} \tag{19}$$

Finally, a solution is obtained for the DeMoivre's law of mortality where $\mu_x(t) = \frac{1}{T-t}$; that is, a uniform distribution of deaths between times t and T. The solution is given by:

$$\begin{aligned} f_a(S,t) = {} & \frac{X e^{pt}}{(q+\lambda)(T-t)}\left(\frac{S}{X e^{pt}}\right) e^{-(q+\lambda)(T-t)} N(-d_1) \\ & - \frac{X e^{pt}}{(r+\lambda-p)(T-t)} e^{-(r+\lambda-p)(T-t)} N(-d_2) \\ & - \left\{\frac{X e^{pt}}{2(q+\lambda)(T-t)}\left[1+\frac{\xi_1}{\sqrt{\xi_1^2+2(q+\lambda)}}\right]\right. \end{aligned}$$

$$
\begin{aligned}
&-\frac{Xe^{pt}}{2(r+\lambda-p)(T-t)}\left[1+\frac{\xi_2}{\sqrt{\xi_2^2+2(r+\lambda-p)}}\right]\Bigg\}\left(\frac{S}{Xe^{pt}}\right)^{m_1}N(-d_3)\\
&+\Bigg\{\frac{Xe^{pt}}{2(r+\lambda-p)(T-t)}\left[1-\frac{\xi_2}{\sqrt{\xi_2^2+2(r+\lambda-p)}}\right]\\
&-\frac{Xe^{pt}}{2(q+\lambda)(T-t)}\left[1-\frac{\xi_1}{\sqrt{\xi_1^2+2(q+\lambda)}}\right]\Bigg\}\left(\frac{S}{Xe^{pt}}\right)^{m_2}N(-d_4)\quad S>Xe^{pt}
\end{aligned}
\tag{20}
$$

and

$$
\begin{aligned}
f_a(S,t)=&\frac{Xe^{pt}\left(1-e^{-(r+\lambda-p)(T-t)}\right)}{(r+\lambda-p)(T-t)}-\frac{Xe^{pt}\left(1-e^{-(q+\lambda)(T-t)}\right)}{(q+\lambda)(T-t)}\left(\frac{S}{Xe^{pt}}\right)\\
&+\frac{Xe^{pt}}{(r+\lambda-p)(T-t)}e^{-(r+\lambda-p)(T-t)}N(d_2)\\
&-\frac{Xe^{pt}}{(q+\lambda)(T-t)}\left(\frac{S}{Xe^{pt}}\right)e^{-(q+\lambda)(T-t)}N(d_1)\\
&+\Bigg\{\frac{Xe^{pt}}{2(q+\lambda)(T-t)}\left[1+\frac{\xi_1}{\sqrt{\xi_1^2+2(q+\lambda)}}\right]\\
&-\frac{Xe^{pt}}{2(r+\lambda-p)(T-t)}\left[1+\frac{\xi_2}{\sqrt{\xi_2^2+2(r+\lambda-p)}}\right]\Bigg\}\left(\frac{S}{Xe^{pt}}\right)^{m_1}N(d_3)\\
&-\Bigg\{\frac{Xe^{pt}}{2(r+\lambda-p)(T-t)}\left[1-\frac{\xi_2}{\sqrt{\xi_2^2+2(r+\lambda-p)}}\right]\\
&-\frac{Xe^{pt}}{2(q+\lambda)(T-t)}\left[1-\frac{\xi_1}{\sqrt{\xi_1^2+2(q+\lambda)}}\right]\Bigg\}\left(\frac{S}{Xe^{pt}}\right)^{m_2}N(d_4)\quad S<Xe^{pt}
\end{aligned}
\tag{21}
$$

The definitions of the ancillary quantities are the same as in the previous case with the exception that the parameter "μ" is replaced by "0".

These results are extended to ratchet GMDB options in Ulm (2014). The PDE is solved using Laplace transforms. The author finds the following solution in the case of a constant force of mortality:

$$
f_a(S,t)=\frac{\mu}{(\mu+r+\lambda)}X-\frac{\mu}{(\mu+q+\lambda)}S+\frac{\mu}{(\mu+r+\lambda)}\frac{X}{m_1-1}\left(\frac{S}{X}\right)^{m_1}\quad S<X
\tag{22}
$$

with m_1 defined as in Eq. (11). As the GMDB is a continuous ratchet, it is not possible for the option to be in the range $S > X$.

As previously noted, the pdf for any arbitrary mortality law can be closely approximated by a sum of exponentials. Ulm (2014) solves the PDE in several additional cases. First, a solution is obtained for the situation where μ is constant until time T, at which point everyone dies. The solution is given by:

$$\begin{aligned} f_a(S,t) = & \frac{\mu}{(\mu+r+\lambda)}X + \frac{r+\lambda}{(\mu+r+\lambda)}Xe^{-(\mu+r+\lambda)(T-t)}N(-d_2) \\ & - \frac{\mu}{(\mu+q+\lambda)}S - \frac{q+\lambda}{(\mu+q+\lambda)}Se^{-(\mu+q+\lambda)(T-t)}N(-d_1) \\ & + \frac{\mu}{r+\mu+\lambda}\frac{X}{m_1-1}\left(\frac{S}{X}\right)^{m_1}N(d_3) + \frac{r+\lambda}{r+\mu+\lambda}\frac{X}{m_2-1}\left(\frac{S}{X}\right)^{m_2}N(d_4) \\ & + \frac{q+\lambda}{(\mu+q+\lambda)}\frac{\sigma^2}{2(r-q)}Se^{-(\mu+q+\lambda)(T-t)}N(d_1) \\ & - \frac{r+\lambda}{(\mu+r+\lambda)}\frac{\sigma^2}{2(r-q)}X\left(\frac{S}{X}\right)^{2\alpha}e^{-(\mu+r+\lambda)(T-t)}N(d_5) \quad S \le X \end{aligned} \tag{23}$$

with definitions

$$\begin{aligned} \alpha &= \frac{1}{2} - \frac{(r-q)}{\sigma^2}; \\ d_1 &= \frac{\ln\left(\frac{S}{X}\right) + \left(r-q+\frac{\sigma^2}{2}\right)(T-t)}{\sigma\sqrt{T-t}}; \\ d_2 &= \frac{\ln\left(\frac{S}{X}\right) + \left(r-q-\frac{\sigma^2}{2}\right)(T-t)}{\sigma\sqrt{T-t}}; \\ d_3 &= \frac{\ln\left(\frac{S}{X}\right) + \sigma^2\sqrt{\xi_1^2 + 2(q+\mu+\lambda)(T-t)}}{\sigma\sqrt{T-t}}; \\ d_4 &= \frac{\ln\left(\frac{S}{X}\right) - \sigma^2\sqrt{\xi_1^2 + 2(q+\mu+\lambda)(T-t)}}{\sigma\sqrt{T-t}}; \\ d_5 &= \frac{\ln\left(\frac{S}{X}\right) + \left(-(r-q)+\frac{\sigma^2}{2}\right)(T-t)}{\sigma\sqrt{(T-t)}}; \end{aligned} \tag{24}$$

Second, a solution is obtained for the situation where μ is constant until time T, at which point the option expires worthless. The solution is given by:

$$f_a(S,t) = \frac{\mu}{(\mu+r+\lambda)}X - \frac{\mu}{(\mu+r+\lambda)}Xe^{-(\mu+r+\lambda)(T-t)}N(-d_2)$$

$$- \frac{\mu}{(\mu + q + \lambda)} S + \frac{\mu}{(\mu + q + \lambda)} S e^{-(\mu+q+\lambda)(T-t)} N(-d_1)$$
$$+ \frac{\mu}{r + \mu + \lambda} \frac{X}{m_1 - 1} \left(\frac{S}{X}\right)^{m_1} N(d_3) + \frac{r + \lambda}{r + \mu + \lambda} \frac{X}{m_2 - 1} \left(\frac{S}{X}\right)^{m_2} N(d_4)$$
$$- \frac{\mu}{(\mu + q + \lambda)} \frac{\sigma^2}{2(r - q)} S e^{-(\mu+q+\lambda)(T-t)} N(d_1)$$
$$+ \frac{\mu}{(\mu + r + \lambda)} \frac{\sigma^2}{2(r - q)} X \left(\frac{S}{X}\right)^{2\alpha} e^{-(\mu+r+\lambda)(T-t)} N(d_5) \quad S \le X \tag{25}$$

Finally, a solution is obtained for the situation where $\mu_x(t) = \frac{1}{T-t}$; that is, a uniform distribution of deaths between times t and T. The solution is given by:

$$f_a(S,t) = \frac{X}{(r+\lambda)(T-t)} - \frac{X e^{-(r+\lambda)(T-t)}}{(r+\lambda)(T-t)} N(-d_2) - \frac{S}{(q+\lambda)(T-t)}$$
$$+ \frac{S e^{-(q+\lambda)(T-t)}}{(q+\lambda)(T-t)} N(-d_1)$$
$$+ \frac{X}{(r+\lambda)(T-t)(m_1-1)} \left(\frac{S}{X}\right)^{m_1} N(d_3) + \frac{X}{(r+\lambda)(T-t)(m_2-1)} \left(\frac{S}{X}\right)^{m_2} N(d_4)$$
$$- \frac{S}{(q+\lambda)(T-t)} \frac{\sigma^2}{2(r-q)} e^{-(q+\lambda)(T-t)} N(d_1)$$
$$+ \frac{X}{(r+\lambda)(T-t)} \frac{\sigma^2}{2(r-q)} \left(\frac{S}{X}\right)^{2\alpha} e^{-(r+\lambda)(T-t)} N(d_5) \quad S \le X \tag{26}$$

The definitions of the ancillary quantities are the same as in the previous case with the exception that the parameter "μ" is replaced by "0".

The mortality laws analyzed to this point are fairly unrealistic and not descriptive of human mortality. Ulm (2014) does make significant progress on the solution of the PDE for the much more realistic Makeham's law of mortality, defined as:

$$\mu(x) = A + Bc^x \tag{27}$$

with parameters A, B, and c set to values that best reproduce empirical mortality distributions. The solution in the at-the-money case is:

$$f_a(1,t) = \frac{A}{r+\lambda+A} + \frac{(r+\lambda)e^{(r+\lambda+A)(t-a)}}{r+\lambda+A} \frac{\Gamma\left(1-(r+\lambda)b, e^{(t-a)/b}\right)}{\Gamma\left(1, e^{(t-a)/b}\right)}$$
$$- \frac{A}{q+\lambda+A} - \frac{(q+\lambda)e^{(q+\lambda+A)(t-a)}}{q+\lambda+A} \frac{\Gamma\left(1-(q+\lambda)b, e^{(t-a)/b}\right)}{\Gamma\left(1, e^{(t-a)/b}\right)}$$
$$+ e^{(r+\lambda+A)(t-a)} e^{e^{(t-a)/b}} \left[\frac{r+\lambda}{r-q} \frac{(1-\alpha)\sigma^2 b}{2} \Gamma\left(-(r+\lambda+A)b, e^{(t-a)/b}\right) \right.$$

$$
\begin{aligned}
&-\frac{q+\lambda}{r-q}\frac{(1-\alpha)\sigma^2 b}{2}\mathrm{e}^{(r-q)(t-a)}\Gamma\left(-(q+\lambda+A)b,\mathrm{e}^{(t-a)/b}\right)\\
&+\frac{2}{\sqrt{\pi}}\mathrm{e}^{-(r+\lambda+A)(t-a)}\frac{1}{\sqrt{\kappa+\gamma}}\int_0^\infty \mathrm{e}^{-u^2}\mathrm{e}^{-\mathrm{e}^{(t-a)/b}\mathrm{e}^{2u^2/\sigma^2(\kappa+\gamma)b}}du\\
&-\frac{2}{\sqrt{\pi}}\frac{q+\lambda}{r-q}\frac{(1-\alpha)\sigma^2 b}{2}\mathrm{e}^{(r-q)(t-a)}\\
&\times\int_0^\infty \mathrm{e}^{-u^2}\Gamma\left(-(q+\lambda+A)b,\mathrm{e}^{(t-a)/b}\mathrm{e}^{2u^2/\sigma^2(1-\alpha)^2 b}\right)du\\
&-\frac{2}{\sqrt{\pi}}\frac{r+\lambda}{r-q}\frac{\alpha\sigma^2 b}{2}\int_0^\infty \mathrm{e}^{-u^2}\Gamma\left(-(r+\lambda+A)b,\mathrm{e}^{(t-a)/b}\mathrm{e}^{2u^2/\sigma^2\alpha^2 b}\right)du\Bigg]
\end{aligned}
\tag{28}
$$

where

$$
\begin{aligned}
&\alpha=\frac{1}{2}-\frac{(r-q)}{\sigma^2};\ \kappa=\frac{2(r+\lambda)}{\sigma^2}+\alpha^2;\ \gamma=\frac{2A}{\sigma^2};\\
&a=-\frac{\ln\left[\frac{B}{\ln(c)}\right]}{\ln(c)};\ b=\frac{1}{\ln(c)};\ m=\frac{B}{\ln(c)};
\end{aligned}
\tag{29}
$$

The integrals clearly converge, but are not available in closed form.

4 The Discounted Density Method

This approach was pioneered by Gerber et al. (2012). They begin by defining the "discounted density function" for the value $X(\tau)$ and running maximum $M(\tau)$ for an exponential stopping time τ distributed with $f_\tau(t)=\mu\mathrm{e}^{-\mu t}$. This function is defined to be:

$$
f^\delta_{X(\tau),M(\tau)}(x,y)=\int_0^\infty \mathrm{e}^{-\delta t}f_{X(t),M(t)}(x,y)\mu\mathrm{e}^{-\mu t}dt \tag{30}
$$

where $f_{x(\tau),M(\tau)}(x,y)$ is the pdf of the value and running maximum at time t. This integral bears a striking resemblance to the density function at an exponential stopping time and can be evaluated using the standard results found, for example, in Borodin and Salminen (2002). In particular, for a Brownian motion with growth rate r,

$$
f^\delta_{X(\tau),M(\tau)}(x,y)=\frac{2\mu}{\sigma^2}\mathrm{e}^{-\alpha x-(\beta-\alpha)y} \tag{31}
$$

where α and β are the negative and positive roots of the quadratic equation:

$$\frac{\sigma^2}{2}x^2 + rx - (\mu + \delta) = 0 \tag{32}$$

All of the options previously analyzed, as well as many others, can be expressed as an expected value of a function of $X(\tau)$ and $M(\tau)$ appropriately discounted. Therefore, one need to only integrate the option function multiplied by the discounted density function to get the GMDB values. Gerber et al. (2012) find values for call options, all-or-nothing call options, put options, all-or-nothing put options, fixed-strike lookback call options, floating strike lookback put options, fractional floating strike lookback put options, fixed-strike lookback put options, floating strike lookback call options, high–low options, up-and-out barrier options, up-and-in barrier options, down-and-out barrier options, and down-and-in barrier options. Most of these options are of theoretical interest only as they are not offered in the GMDB market. The formulas are not reproduced here due to space considerations, but they can be found in the original paper. They are also able to obtain formulas for options that expire at time T as well as for DeMoivre's law mortality.

Gerber et al. (2013) extend these results to situations where the fund process follows a jump-diffusion Lévy process. They are again able to use results on the exponential stopping times of such processes to get closed-form solutions for option prices when the jump sizes are exponentially distributed. They also suggest that "knock-out" options are a reasonable way to model lapses. This assumes that when the option is sufficiently out of the money the individual will surrender the policy, which is the equivalent of the option expiring worthless. As before, the large number of solutions found in Gerber et al. (2013) will not be reproduced here, and the interested reader is referred to the original paper.

Siu et al. (2015) extend these results to GMDB options where the fund follows a regime-switching double-exponential jump-diffusion process. They are able to obtain closed-form expressions for the Laplace transforms of the option values, which are then inverted numerically.

5 Extensions of the Analytic Methods to Related Problems

It is sometimes the case in insurance and risk management problems that one is not only interested in the expected value of a quantity but also the full distribution or percentiles of the option outcomes. For instance, if an insurance company wishes to be, say, 99% certain of avoiding bankruptcy then the 99th percentile of the distribution of outcomes needs to be determined. The PDE approach has been used for these types of problems, but not many analytic results are known. The interested reader is directed to Feng and Volkmer (2012) and Feng and Huang (2016) for details.

Variable annuity contracts can contain other types of riders in addition to guaranteed minimum death benefits. These include guaranteed minimum accumulation benefits, guaranteed minimum income benefits, guaranteed minimum withdrawal benefits, and guaranteed lifetime withdrawal benefits, often identified collectively with GMDBs as GMxBs. The interested reader is referred to Bauer et al. (2008) or the books by Hardy (2003) or Feng (2018) for descriptions of these riders and possible methods for determining the option values. Some analytic solutions can be found in Feng and Volkmer (2016) and Feng and Jing (2017).

6 Conclusions and Future Research Directions

This paper reviews work to date on analytic solutions for GMDB options embedded in variable annuity contracts. It presents the common solution methods including direct integration, PDE methods, and discounted density approaches. Analytic solutions have been obtained to most common GMDB options (as well as many uncommon ones) when the force of mortality is constant. Only a small number of analytic solutions have been obtained for more realistic mortality laws, and there is scope for work in this area. Methodologies have been developed for determining quantiles and percentiles of the option distribution but very few analytic results have been obtained with room for further research in this area. Finally, there are almost no analytic solutions for other types of GMxB options, and there is space in this area for further research. Table 1 summarizes the strong points and limitations of the methods considered in this review.

Table 1 Strong points and limitations of the methods considered in this review

Method	Papers	Strong features	Limitations
Direct integration	Milevsky and Posner (2001)	Easy to understand	Only used for constant force of mortality Only used for at-the-money options Only used for geometric Brownian motion
PDE methods	Milevsky and Salisbury (2001) Ulm (2006) Ulm (2008) Ulm (2014)	Used for many types of mortality laws Implied policyholder options can be analyzed	Complicated to apply in practice Only used for geometric Brownian motion
Discounted density methods	Gerber et al. (2012) Gerber et al. (2013) Siu et al. (2015)	Solutions can be obtained for many option types Used for processes other than geometric Brownian motion	Only used for constant force of mortality and DeMoivre's law

References

Bauer D, Kling A, Russ J (2008) A universal pricing framework for guaranteed minimum benefits in variable annuities. ASTIN Bull 38(2):621–651

Borodin AN, Salminen P (2002) Handbook of Brownian motion—facts and formulae, 2nd edn. Birkhäuser Verlag, Basel, Switzerland

Dufresne D (2007) Fitting combinations of exponentials to probability distributions. Appl Stoch Model Bus Ind 23(1):23–48

Feng R (2018) An introduction to computational risk management of equity-linked insurance. CRC, Boca Raton, FL

Feng R, Volkmer HW (2016) An Identity of hitting times and its application to the valuation of guaranteed minimum withdrawal benefit. Math Financ Econ 10(2):127–149

Feng R, Huang H (2016) Statutory financial reporting for variable annuity guaranteed death benefits: market practice, mathematical modeling and computation. Insur: Math Econ 67:54–64

Feng R, Jing X Analytical valuation and hedging of variable annuity guaranteed lifetime withdrawal benefits. Insur: Math Econ 72:36–48

Feng R, Volkmer HW (2012) Analytical calculation of risk measures for variable annuity guaranteed benefits. Insur: Math Econ 51(3):636–648

Gao J, Ulm ER (2018) Optimal consumption and allocation in variable annuities with guaranteed minimum death benefits. Insur: Math Econ 51(3):586–598

Gao J, Ulm ER (2015) Optimal allocation and consumption with guaranteed minimum death benefits, external income and term life insurance. Insur: Math Econ 61:87–98

Gerber HU, Shiu ESW, Yang H (2012) Valuing equity-linked death benefits and other contingent options: a discounted density approach. Insur: Math Econ 51:73–92

Gerber HU, Shiu ESW, Yang H (2013) Valuing equity-linked death benefits in jump diffusion models. Insur: Math Econ 53:615–623

Goldman B, Sosin H, Gatto ME (1979) Path dependent options: buy at the low, sell at the high. J Finance 34:1111–1127

Hardy M (2003) Investment guarantees. Wiley, Hoboken, N.J.

Haug EG (2007) The complete guide to option pricing formulas, 2nd edn. McGraw-Hill, New York

Milevsky M (2006) The calculus of retirement income. Cambridge University Press, New York

Milevsky M, Posner SE (2001) The titanic option: valuation of the guaranteed minimum death benefit in variable annuities and mutual funds. J Risk Insur 68(1):93–128

Milevsky MA, Salisbury TS (2001) The real option to lapse and the valuation of death-protected investments. In: Conference proceedings of the 11th annual international AFIR colloquium

Moenig T (2012) Optimal policyholder behavior in personal savings products and its impact on valuation. Risk Management and Insurance Dissertations 28, Georgia State University

Moenig T, Zhu N (2018) Lapse-and-reentry in variable annuities. J Risk Insur 85(4):911–938

Siu CC, Yam SCP, Yang H (2015) Valuing equity-linked death benefits in a regime-switching framework. ASTIN Bull 45(2):355–395

Ulm ER (2006) The effect of the real option to transfer on the value of guaranteed minimum death benefits. J Risk Insur 73(1):43–69

Ulm ER (2008) Analytic solution for return of premium and rollup guaranteed minimum death benefit options under some simple mortality laws. ASTIN Bull 38(2):543–563

Ulm ER (2014) Analytic solution for Ratchet guaranteed minimum death benefit options under a variety of mortality laws. Insur: Math Econ 58:14–23

Mathematical Modeling and Inverse Problem Approaches for Functional Clothing Design Based on Thermal Mechanism

Dinghua Xu and Tingyue Li

1 Background of the IPTMD

Textile materials, like many other materials, should be designed and engineered to possess specific attributes or properties for different end-user (Xu 2014; Huang 2008). People have the common opinion that *The better clothing, the happier life; the better clothing, the less thermal injury; the smarter clothing, the more upgraded industries*. Compared to other materials, such as metal, plastics, and electronics, engineering of textile materials presents a much greater challenge, owing to the lack of the precise relationships between material–processing–structural variables on one hand and properties/performance on the other hand as well as the nonlinear nature of the multivariable interactions.

This is however changing, thanks to the efforts of generations of textile scientists. Within limits, we can now predict the properties and performance of textile materials by applying theoretical models and empirical relations. Moreover, it would be further desirable to quantify the material, processing, and structural parameters from the anticipated end-use requirements. This leads to a class of inverse problems in mathematics, which makes the mathematical research for the textile industry so interesting and valuable.

Although the end-use requirements of textile materials include durability, comfort, protection, tailorability, appearance, etc., the thermal comfort and thermal safety are the two important factors among all of them. So this paper mainly focuses on the two kinds of functional clothing, i.e., thermal comfort clothing (TCC) and thermal protective clothing (TPC). The inverse problem of textile materials in terms of thermal

D. Xu
Department of Mathematics, College of Sciences, Zhejiang Sci-Tech University, Hangzhou 310018, Zhejiang Province, People's Republic of China

D. Xu (✉) · T. Li
School of Mathematics, Shanghai University of Finance and Economics, Shanghai 200433, People's Republic of China
e-mail: dhxu6708@zstu.edu.cn

J. Cheng et al. (eds.), *Proceedings of the Forum "Math-for-Industry" 2018*,
Mathematics for Industry 35, https://doi.org/10.1007/978-981-16-5576-0_4

comfort, particularly thermo-physiological comfort, is a good starting point, as about 50% of textile materials are used for apparel and comfort is the primary concern of clothing. The thermal protective clothing is essentially an important garment to protect the human body from thermal damage. Therefore, the TCC and the TPC design should be considered so that the physical and structural parameters of textiles are determined to satisfy the thermal comfort index or reduce the thermal injury.

So far many researchers have treated these partial differential equations as direct problems to find solutions with respect to specific parameters of the human body—clothing-environment system through different approaches including finite difference method, finite volume method, finite element method and volume–time–domain recursive method (Fan et al. 2000, 2004; Li et al. 2004; Fan and Wei 2002; Wu and Fan 2008; Ye et al. 2008; Du et al. 2009). Few have treated it as an inverse problem for the TCC/TPC design, viz. to find the desirable material parameters from the constraints and end-use requirements. The inverse problem of textile material design (IPTMD) should be professionally treated by mathematicians (Engl et al. 1998; Stuart 2010).

Generally speaking, inverse problems are mathematically ill-posed; that is, their solutions may not satisfy the requirements of existence, uniqueness or stability. We are so pleased to see a growing number of mathematicians interested in this problem. We are witnessing the challenge of deriving well-posed IPTMD through stabilization algorithms and investigated various numerical methods for the solution. These methods include regularization methods, quasi-solution methods, direct search methods, iterative algorithms, and stochastic algorithms (Xu et al. 2010, 2011, 2012, 2013, 2014, 2018; Xu 2014; Chen et al. 2011; Xu and Ge 2012; Xu and Wen 2014; Xu et al. 2015; Yu and Xu 2015; Yu et al. 2015a, b; Xu and Cui 2016; Ge et al. 2017).

The TCC and TPC design will be mathematically studied by means of heat and moisture transfer law within the body–clothing–environment system. The mathematical model of the heat and moisture transfer can be deduced to partial differential equations or fractional-order partial differential equations(PDEs) by means of the thermal mechanism (Xu et al. 2012, 2014; Xu and Ge 2012; Xu and Wen 2014; Xu 2014; Yu and Xu 2015; Yu et al. 2015a, b; Ge et al. 2017). The reason why we reformulate TCC design based on PDEs model is motivated by classic heat and moisture transfer process; see Sect. 2. Meanwhile, the reason why we reformulate TPC design based on fractional PDEs model is motivated by superdiffusion characteristics under the high environmental temperature; see Sect. 3.

Henceforth, the study in the paper belongs to the direct problems and inverse problems for partial differential equations (Friedman 1964; Yosida 1999). In this paper, we progressively make a review on the inverse problems for the TCC and TPC design together with the mathematical model of the direct problems. The inverse problems will be classified as the determination of thickness, thermal conductivity, and porosity of the clothing system. The techniques developed in this work can be applied to more complex material designs whether in textiles or other areas. We believe that this is just the start of fruitful researches in this direction, and it will be a promising topic in the industrially intelligent manufacturing.

Nomenclature

$C_a(x,t)$	water vapor concentration in the inter-fiber void space (kg m^{-3})
$C_a^*(x,t)$	saturated water vapor concentration in the inter-fiber void space (kg m^{-3})
$C_e(t)$	water vapor concentration between outer covering fabrics and surrounding (kg m^{-3})
$C_1(t)$	water vapor concentration in outer covering fabrics (kg m^{-3})
$C_0(x)$	initial water vapor concentration in batting (kg m^{-3})
C_v	effective volumetric heat capacity of the fibrous batting (kJ m^{-3} K^{-1})
D_a	diffusion coefficient of water vapor in the air (m^2 s^{-1})
$F_L(x,t)$	total heat radiation incident traveling to the left (kJ m^{-2} s^{-1})
$F_R(x,t)$	total heat radiation incident traveling to the right (kJ m^{-2} s^{-1})
$RH(x,t)$	relative humidity of the surroundings (%)
$RH_b(t)$	relative humidity in the microclimate area of TCC (%)
$RH_c(t)$	relative humidity in the outer environment of TCC (%)
$T(x,t)$	temperature in fabrics (K or °C)
$T_b(t)$	temperature in the microclimate area of TCC (K or °C)
$T_e(t)$	temperature in the outer environment of TCC (K or °C)
$T_0(t)$	temperature in the outer environment of TPC (K or °C)
$T_1(t)$	temperature in the human body (K or °C)
$T_I(x)$	initial temperature in batting (K or °C)
$\Gamma(x,t)$	total rate of (de)sorption, condensation, freezing and/or evaporation (kg m^{-3}s^{-1})
ξ, ξ_i	surface emissivity of the inner and outer covering fabrics (i = 1:inner fabric; i = 2:outer fabric)
k, k_i	effective heat conductivity of the fibrous batting (i=1:inner batting, i = 2:outer batting) and in microclimate area respectively (kJ m^{-1} K^{-1} s^{-1})
κ_γ	thermal conductivity of textile (kJ m^{-1} K^{-1} s^{-1})
λ	latent heat of (de)sorption of fibers or condensation of water vapor (kJ kg^{-1})
β	radiative sorption constant of the fibers (m^{-1})
σ	Boltzmann constant (kJ m^{-2} K^{-4} s^{-1})
$\varepsilon, \varepsilon_i$	porosity of the fibrous batting (i=1:inner fabric; i=2:outer fabric)
τ	effective tortuosity of the fibrous batting
w_1	water vapor resistance of inner and outer fabrics
h_c	convective vapor transfer coefficient (m s^{-1})
h_i	constants dependent on Stefan–Boltzmann constant and the emissivity of contiguous objects on the contact interface ($i = 1$: the contact interface between the fabric and the body (W m^{-2}K^{-4}) ; $i = 2$: the contact interface between the cold air and the fabric)
p_i	heat transfer coefficient ($i = 1$: between the fabric and the body ; $i = 2$: between the environment and the fabric).

2 Mathematical Model of Dynamic Heat–Moisture Transfer within the TCC System

Heat–Moisture Transfer Model within the Textiles We consider a clothing model consisting of a thin inner fabric layer, a thick fibrous batting, and a thin outer fabric in the body–clothing–environment system; see Fig. 1. The thin inner fabric is close to human skin and the thin outer fabric is next to the outer environment. We assume that the fibrous batting is isotropic. See Xu (2014); Fan et al. (2004); Xu and Ge (2012) for detailed information.

Based on the conservation of heat energy and mass balance, the governed equations can be described in the form of coupled equations

$$\begin{cases} C_v(x,t)\frac{\partial T}{\partial t} = \frac{\partial}{\partial x}(k(x,t)\frac{\partial T}{\partial x}) + \frac{\partial F_{\mathrm{L}}}{\partial x} - \frac{\partial F_{\mathrm{R}}}{\partial x} + \lambda(x,t)\Gamma(x,t), \\ \frac{\partial F_{\mathrm{L}}}{\partial x} = \beta F_{\mathrm{L}} - \beta\sigma T^4(x,t), \\ \frac{\partial F_{\mathrm{R}}}{\partial x} = -\beta F_{\mathrm{R}} + \beta\sigma T^4(x,t), \quad (x,t) \in \Omega_T = (0,L) \times (t_1,t_2). \end{cases} \tag{1}$$

All kinds of heat and moisture transfer processes, such as heat conduction, radiation, and sorption flow, are considered in the single layer porous fabric batting in (1). The first equation describes a dynamic model of heat and moisture transfer with sorption and condensation rate $\Gamma(x,t)$ in porous fabric. The second and third equations describe the attenuation of the left radiation flux $F_{\mathrm{L}}(x,t)$ and the right radiation flux $F_{\mathrm{R}}(x,t)$. The dynamic changes in temperature, moisture concentration, (de)sorption and thermal radiation as well as the effect of water content on the effective thermal conductivity are considered in the heat transfer process.

The water vapor concentration in the textile can be described as follows:

$$\varepsilon\frac{\partial C_a}{\partial t} = \frac{D_a\varepsilon}{\tau}\frac{\partial^2 C_a}{\partial x^2} - \Gamma(x,t), \quad (x,t) \in \Omega_T. \tag{2}$$

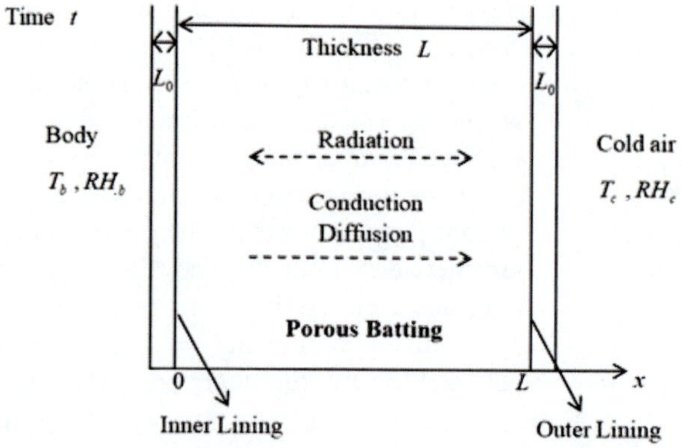

Fig. 1 Schematic diagram of the body–TCC–environment system

According to mass conservation, Eq. 2 is derived by the mass balance relationship for the moisture accumulation.

When the relative humidity reaches 100% or above, condensation or freezing takes place in fabric and additionally liquid water occurs. The vapor concentration can be saturated and solely determined by the temperature

$$\Gamma(x,t) = \left(\frac{D_a}{\tau}\frac{\partial^2 C_a^*(x,t)}{\partial x^2} - \frac{\partial C_a^*(x,t)}{\partial t}\right), \quad (x,t) \in \Omega_T, \tag{3}$$

where the saturated water vapor concentration $C_a^*(x,t)$ can be given by

$$C_a^*(x,t) = 216.5 \times Vap(T(x,t)) \times 10^{-6}/T(x,t), \quad (x,t) \in \Omega_T, \tag{4}$$

$$\text{Vap}(T) = \begin{cases} 1013.25\mathrm{e}^{13.3185s - 1.976s^2 - 0.6445s^3 - 0.1299s^4}, & T \leqslant 273.15 \\ 10^{10.5380997 - 2663.91/T}, & T > 273.15. \end{cases} \tag{5}$$

where $s = T - 273.15$. The relative humidity RH(x,t) in the interfiber space can be determined by the relationship

$$\text{RH}(x,t) = \frac{T(x,t) \times C_a(x,t) \times 10^6}{216.5 \times \text{Vap}(T(x,t))}, \quad (x,t) \in \Omega_T, \tag{6}$$

where the vapor pressure Vap(T) can be obtained from (5).

The dynamic heat and mass transfer model can be mathematically formulated into the following initial and boundary value problem:

$$\begin{cases} \begin{cases} C_{\mathrm{v}}(x,t)\frac{\partial T}{\partial t} = \frac{\partial}{\partial x}(k(x,t)\frac{\partial T}{\partial x}) + \frac{\partial F_{\mathrm{L}}}{\partial x} - \frac{\partial F_{\mathrm{R}}}{\partial x} + \lambda(x,t)\Gamma(x,t), \\ \frac{\partial F_{\mathrm{L}}}{\partial x} = \beta F_{\mathrm{L}} - \beta\sigma T^4(x,t), \\ \frac{\partial F_{\mathrm{R}}}{\partial x} = -\beta F_{\mathrm{R}} + \beta\sigma T^4(x,t), \quad (x,t) \in \Omega_T; \end{cases} \\ \begin{cases} T(x,t_1) = T_{\mathrm{I}}(x), \quad x \in (0,L), \\ -k\frac{\partial}{\partial x}T(0,t) = p_1(T_b(t) - T(0,t)), \\ k\frac{\partial}{\partial x}T(L,t) = p_2(T_c(t) - T(L,t)), \quad t \in (t_1,t_2); \end{cases} \\ \begin{cases} (1-\xi_1)F_{\mathrm{L}}(0,t) + \xi_1\sigma T^4(0,t) = F_{\mathrm{R}}(0,t), \\ (1-\xi_2)F_{\mathrm{R}}(L,t) + \xi_2\sigma T^4(L,t) = F_{\mathrm{L}}(L,t), \quad t \in (t_1,t_2). \end{cases} \end{cases} \tag{\textbf{DP1}}$$

Meanwhile, we consider the moisture transfer Eq. (2) with initial and boundary conditions

$$\begin{cases} \varepsilon\frac{\partial C_a}{\partial t} = \frac{D_a\varepsilon}{\tau}\frac{\partial^2 C_a}{\partial x^2} - \Gamma(x,t), \quad (x,t) \in \Omega_T, \\ C_a(x,t_1) = C_0(x), \quad x \in (0,L), \\ C_a(L,t) = C_e(t), \ \frac{D_a\varepsilon}{\tau}\frac{\partial C_a}{\partial x}\Big|_{x=L} = \frac{C_1(t) - C_a|_{x=L}}{w_1 + (1/h_c)}, \quad t \in (t_1,t_2). \end{cases} \tag{\textbf{SP}}$$

The above heat transfer problem is called *direct problem* (in abbreviation **DP1**). The water vapor transfer problem is called *sideways problem* (in abbreviation **SP**).

Denote

$$c_1 = \frac{\xi_1}{\beta} T^4(0,t) - (1-\xi_1)\, c_2,$$

$$c_2 = \frac{1}{(1-\xi_2)\,\beta\,(1-\xi_1)\,\mathrm{e}^{-\beta L} - \beta \mathrm{e}^{\beta L}} \Bigg[(1-\xi_2)\,\beta e^{-\beta L} \int_0^L \mathrm{e}^{\beta x} T^4(x,t)\mathrm{d}x$$

$$+ \beta \mathrm{e}^{\beta L} \int_0^L \mathrm{e}^{-\beta x} T^4(x,t)\mathrm{d}x + (1-\xi_2)\,\xi_1 \mathrm{e}^{-\beta L} T^4(0,t) + \xi_2 T^4(L,t) \Bigg].$$

In the **DP1**, based on both the second and the seventh equations, we can derive that

$$F_{\mathrm{L}}(x,t) = -\beta\sigma \mathrm{e}^{\beta x} \left[\int_0^x \mathrm{e}^{-\beta y} T^4(y,t)\mathrm{d}y + c_2 \right]. \tag{7}$$

Similarly, both the third and the eighth equations give that

$$F_{\mathrm{R}}(x,t) = \beta\sigma \mathrm{e}^{-\beta x} \left[\int_0^x \mathrm{e}^{\beta y} T^4(y,t)\mathrm{d}y + c_1 \right] - 2\beta\sigma T^4(x,t). \tag{8}$$

In the **DP1**, we will find $T(x,t)$, $(x,t) \in \Omega_T$, and meanwhile in the **SP** we will find $C_a(x,t)$, $x \in [0,L)$, $t \in (t_1,t_2)$.

Remark 1 (Existence, uniqueness, and stability of the solution to the direct problems) The well-posedness results for the **DP1** can be referred to the papers, for example, Xu et al. (2014), Yu and Xu (2015).

Remark 2 (Numerical computation for the sideways problems) The stabilized numerical algorithms need developing for the **SP**. The numerical results and numerical examples can be found in Yu et al. (2015a, b).

Remark 3 (Model of heat–moisture transfer within multilayered textiles) A variety of models can be derived to describe the heat–moisture transfer within multilayered textiles, for example, we can refer to Xu et al. (2014), Xu (2014).

Remark 4 (Model of heat radiation transfer) If the heat transfer within the textiles is considered only by heat conductivity and heat radiation, another model can be formulated as follows:

$$\begin{cases} C_v(x,t)\frac{\partial T}{\partial t} = \frac{\partial}{\partial x}(k(x,t)\frac{\partial T}{\partial x}), & (x,t) \in \Omega_T, \\ T(x,t_1) = T_I(x), & x \in (0,L), \\ -k\frac{\partial}{\partial x}T(0,t) = h_1(T_b^4(t) - T^4(0,t)), & t \in (t_1,t_2), \\ k\frac{\partial}{\partial x}T(L,t) = h_2(T_c^4(t) - T^4(L,t)), & t \in (t_1,t_2). \end{cases} \tag{\textbf{DP2}}$$

The **DP2** can be seen as a simplification of the **DP1**, and its well-posedness results were derived via the theorem in the paper (Yang et al. 2008), where a heat transfer model in composite materials with Stefan–Boltzmann interface conditions was presented, and the global uniqueness result was derived.

3 Mathematical Model of Dynamic Heat Transfer within the TPC System

3.1 Fractional Description for Superdiffusion

We consider the heat transfer in firefighter protective clothing during a flash fire exposure (Chitrphiromstri and Kuznetsov 2005; Song et al. 2008; Elgafy and Mishra 2014). The practical experience has taught us that this process will be very different from the case under low temperature, viz. the high heat and moisture make the transmission process much faster than the classical case under low temperature.

In recent years, anomalous diffusion which deviates the classical Fickian diffusion has gained considerable attention, due mainly to its successful applications in science and engineering (Metzler and Klafter 2000; Klafter and Sokolov 2005; Metzler and Klafter 2004). Anomalous diffusion is characterized through the power law form

$$\langle x^2(t) \rangle \sim t^\alpha, \quad \alpha \neq 1, \tag{9}$$

and can be modeled by fractional partial differential equations, where $\langle x^2(t) \rangle$ denotes the mean squared displacement. According to the value of the diffusion exponent α, transport process is distinguished as following relationship

$$\langle x^2(t) \rangle \sim t^\alpha \begin{cases} \text{subdiffusion/dispersive}, & 0 < \alpha < 1, \\ \text{normal diffusion}, & \alpha = 1, \\ \text{superdiffusion}, & \alpha > 1. \end{cases}$$

Inspired by the model proposed by Fan et al. (2000), Fan and Wei (2002), Fan et al. (2004), Wu and Fan (2008), Li et al. (2004), and the faster transmission of superdiffusion (Metzler and Klafter 2000; Metzler et al. 1998), we present a spatial fractional heat transfer model to describe the faster transmission process instead of the classical Fourier's law. Fan's model has been considered extensively in Du et al. (2009), Fan et al. (2004), Xu and Ge (2012), Xu (2014), Xu et al. (2013), Xu et al. (2014), but yet we haven't found the corresponding spatial fractional model. In the numerical simulation, the probable cause will be given; that is, the fractional model goes against the real situation under low temperature because of the faster transmission.

Figure 2 shows the schematic view of the body–TPC–environment system.

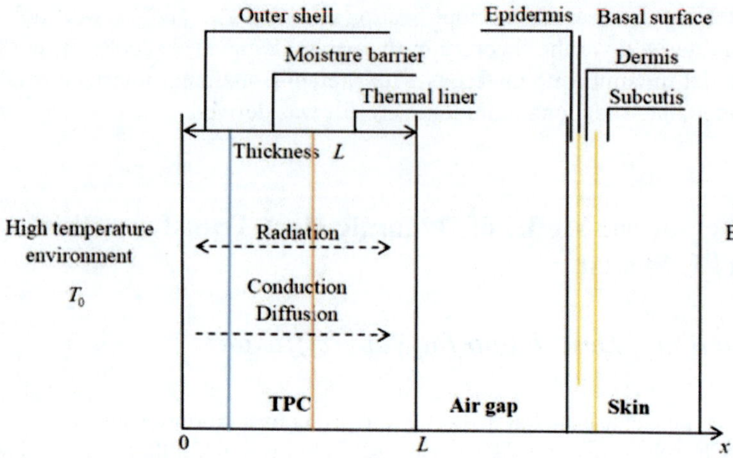

Fig. 2 Schematic diagram of the body–TPC–environment system

As illustrated in the introduction, the faster transmission process of the situation forces us to introduce an extra superdiffusion assumption: Heat conduction within the porous batting is non-Fourier and can be described by the superdiffusion model.

According to continuous time random walk (CTRW) scheme, the following standard fractional differential equation is derived

$$\frac{\partial W}{\partial t} = K^{\mu}{}_{-\infty}D_x^{\mu} W(x,t) \qquad (1 < \mu < 2), \tag{10}$$

where $W(x,t)$ is the pdf of being at a certain position x at time t (called the propagator) and ${}_{-\infty}D_x^{\mu}$ is Weyl operator which in one dimension is equivalent to the Riesz operator ∇^{μ} (Metzler and Klafter 2000; Metzler et al. 1998); K^{μ} is a physical constant.

Different from the relationship (10), the Weyl operator ${}_{-\infty}D_x^{\mu}$ is replaced by utilizing the Riemann–Liouville differential operator D_{0+}^{γ} $(1 < \gamma < 2)$ in the following equations because we only consider the problem in a finite domain. The Riemann–Liouville fractional derivative of order γ $(1 < \gamma \leqslant 2)$ is defined by

$$D_{0+}^{\gamma}u(x,t) = \begin{cases} \frac{1}{\Gamma(2-\gamma)} \frac{\partial^2}{\partial x^2} \int\limits_0^x \frac{u(s,t)}{(x-s)^{\gamma-1}}\, ds, & 1 < \gamma < 2, \\ \frac{\partial^2 u}{\partial x^2}, & \gamma = 2. \end{cases}$$

In the numerical simulation of the fifth section, we observe that it is more appropriate to choose γ such that $1.5 < \gamma < 2$ in the fractional differential equation, and hence, the heat transfer in firefighter protective clothing satisfies the superdiffusion law.

3.2 Mathematical Model for TPC System

Based on the physical and mathematical consideration, the temperature $T(x,t)$ and heat radiation $F_{\mathrm{L}}(x,t)$, $F_{\mathrm{R}}(x,t)$ in firefighter protective clothing satisfy the following partial differential equations

$$\begin{cases} C_v \frac{\partial T}{\partial t}(x,t) = \underbrace{\kappa_\gamma (D_{0+}^{\gamma} T)(x,t)}_{\text{heat conduction}} + \underbrace{\frac{\partial F_{\mathrm{L}}(x,t)}{\partial x} - \frac{\partial F_{\mathrm{R}}(x,t)}{\partial x}}_{\text{heat radiation}} + \underbrace{\lambda \Gamma(x,t)}_{\text{phase change}}, \\ \qquad (x,t) \in \Omega \times (0, t_{\mathrm{f}}), \\ \frac{\partial F_{\mathrm{L}}(x,t)}{\partial x} = \beta F_{\mathrm{L}}(x,t) - \beta \sigma T^4(x,t), \\ \frac{\partial F_{\mathrm{R}}(x,t)}{\partial x} = -\beta F_{\mathrm{R}}(x,t) + \beta \sigma T^4(x,t), \end{cases} \tag{11}$$

together with the initial condition

$$T(x,0) = T_I(x), \quad x \in \overline{\Omega}, \tag{12}$$

the left boundary value conditions

$$\begin{cases} -\kappa_\gamma \frac{\partial}{\partial x} T(0,t) = p_2(T_0(t) - T(0,t)), \quad t \in [0, t_{\mathrm{f}}], \\ (1-\xi_1) F_{\mathrm{L}}(0,t) + \xi_1 \sigma T^4(0,t) = F_{\mathrm{R}}(0,t), \end{cases} \tag{13}$$

and the right boundary value conditions

$$\begin{cases} \kappa_\gamma \frac{\partial}{\partial x} T(L,t) = p_1(T_1(t) - T(L,t)), \quad t \in [0, t_{\mathrm{f}}], \\ (1-\xi_2) F_{\mathrm{R}}(L,t) + \xi_2 \sigma T^4(L,t) = F_{\mathrm{L}}(L,t), \end{cases} \tag{14}$$

where $\Omega = (0, L)$, L is the thickness of firefighter protective clothing, t_{f} is a preestablished time.

The above heat transfer problem (11)–(14) is called *direct problem* (in abbreviation **DP3**). Similar to (7)–(8), the descriptions of $F_{\mathrm{L}}(x,t)$ and $F_{\mathrm{R}}(x,t)$ in (11) can be explicitly derived. In the **DP3**, we will find the temperature distribution $T(x,t)$, $(x,t) \in \Omega \times [0, t_{\mathrm{f}}]$.

Remark 5 It is obvious that the classical Fourier's law is replaced by the fractional second constitutive relation

$$q(x,t) = -\kappa_\gamma \left(D_{0+}^{\gamma-1} T \right)(x,t), \quad 1 < \gamma < 2,$$

where $q(x,t)$ is the heat flux due to conduction.

Remark 6 Chitrphiromstri and Kuznetsov (2005) presented a model of heat and moisture transport in firefighter protective clothing during a flash fire exposure, where heat radiation is modeled by Beer's radiation attenuation model

$$q_{\mathrm{rad}}(x) = q_{\mathrm{rad}}(0)\mathrm{e}^{-\alpha x}.$$

Here, $q_{\mathrm{rad}}(x)$ is the incident radiation heat flux from the flame onto the fabric and α is the extinction coefficient of the fabric. In contrast, the heat radiation in Fan's model is simulated by two flux approximation. The two flux approximation is not limited to the optically thin or optically thick approximations, and it is considered as the appropriate technique for the very thin fibrous insulation spacers in applications. The two flux approximation of the heat radiation modeled by Stefan–Boltzmann law is given by

$$\begin{cases} \frac{\partial F_{\mathrm{L}}}{\partial x} = \beta F_{\mathrm{L}} - \beta\sigma T^4(x,t), \\ \frac{\partial F_{\mathrm{R}}}{\partial x} = -\beta F_{\mathrm{R}} + \beta\sigma T^4(x,t) \end{cases}$$

with radiation boundary conditions

$$\begin{cases} (1-\xi_1)F_{\mathrm{L}}(0,t) + \xi_1\sigma T^4(0,t) = F_{\mathrm{R}}(0,t), & 0 < t < t_{\mathrm{f}}, \\ (1-\xi_2)F_{\mathrm{R}}(L,t) + \xi_2\sigma T^4(L,t) = F_{\mathrm{L}}(L,t), & 0 < t < t_{\mathrm{f}}. \end{cases}$$

The radiation conditions stand for the radiative heat transfer at the interface between the inner thin fabric and the fibrous batting and that between the outer thin fabric and the fibrous batting (Fan et al. 2004).

Remark 7 $\lambda\Gamma(x,t)$ in (11) describes the phase change in heat and moisture transfer, which is very complicated in Fan's model. In Fan's model, $\lambda\Gamma(x,t)$ is determined by an empirical equation under low temperature, which may not be valid for the situation we concern. In Chitrphiromstri and Kuznetsov (2005), the phase change is modeled by solid-phase continuity equation and gas-phase diffusivity equation. This thermodynamic process is much more complicated than Fan's model and leads to strong coupling of heat and moisture.

Remark 8 The unique existence and conditional stability of the weak solution to the fractional heat transfer model can be derived by the PDE theory. One can refer to the paper (Yu et al. 2016). Multilayered TPC models can be referred to the paper such as Yang et al. (2008), Xu et al. (2014). Various treatment of PDEs and boundary conditions can be referred to the paper, for example Ye et al. (2008), Podlubny (1999), Yu et al. (2016), Ervin and Roop (2006), Jin et al. (2015), Hua and Yu (2013), Du et al. (2012), Bowles and Agueh (2015), where the theoretical analysis and numerical implementation can be found for our consideration.

4 Numerical Computation for the TPC Direct Problems to Determine the Fractional Order

4.1 Numerical Algorithm

We adopt the shifted Grünwald formula at all time levels for approximating the fractional derivative (Liu et al. 2005)

$$D_{0+}^{\gamma}T(x_i, t_{n+1}) = \frac{1}{h^{\gamma}}\sum_{j=0}^{i+1} g_j T\left(x_i - (j-1)h,\ t_{n+1}\right) + O(h).$$

Here the normalized Grünwald weights are defined by

$$g_0 = 1,\ g_j = (-1)^j \frac{\gamma(\gamma-1)\ldots(\gamma-j+1)}{j!},\quad j = 1, 2, 3, \ldots.$$

Particularly, $g_0 = 1,\ g_1 = -\gamma,\ g_2 = \frac{\gamma(\gamma-1)}{2}$. Based on the equation of (7)-(8), we can derive the formulation of $\frac{\partial F_L}{\partial x},\ \frac{\partial F_R}{\partial x}$. Denote

$$\begin{aligned}\Theta(T(x,t)) &= \frac{\partial F_L}{\partial x} - \frac{\partial F_R}{\partial x} + \lambda(x,t)\Gamma(x,t)\\ &= -\beta^2\sigma e^{\beta x}\left[\int_0^x e^{-\beta y}T^4(y,t)dy + c_2\right] + \beta^2\sigma e^{-\beta x}\left[\int_0^x e^{\beta y}T^4(y,t)dy + c_1\right]\\ &\quad -2\beta\sigma T^4(x,t) + \lambda(x,t)\Gamma(x,t),\end{aligned}$$

and $T_i^n,\ \Theta_i^n$ by $T(x_i, t_n),\ \Theta(T(x_i, t_n))$ approximately, respectively; thus, we have

$$\begin{aligned}C_v\frac{T_i^{n+1}-T_i^n}{\Delta t} &= \kappa_\gamma\frac{1}{h^\gamma}\sum_{j=0}^{i+1} g_j T\left(x_i-(j-1)h,\ t_{n+1}\right) + \Theta_i^{n+1}\\ &= \kappa_\gamma\frac{1}{h^\gamma}\sum_{j=0}^{i+1} g_j T_{i+1-j}^{n+1} + \Theta_i^{n+1}\quad,\\ &\quad i = 1, 2, \ldots, M-1,\quad n = 0, 1, \ldots, N-1.\end{aligned}$$

Let $s = \frac{\kappa_\gamma \Delta t}{C_v h^\gamma},\ r = \frac{\tau}{C_v}$. We have

$$T_i^{n+1} - s\left(g_0T_{i+1}^{n+1} + g_1T_i^{n+1} + \cdots + g_{i+1}T_0^{n+1}\right) = T_i^n + r\Theta_i^{n+1}$$

or

$$\begin{aligned}&(1-sg_1)T_1^{n+1} - sg_0T_2^{n+1} = T_1^n + r\Theta_1^{n+1} + sg_2T_0^{n+1},\\ &-sg_iT_1^{n+1} - sg_{i-1}T_2^{n+1} - \cdots - sg_2T_{i-1}^{n+1} + (1-sg_1)T_i^{n+1} - sg_0T_{i+1}^{n+1}\\ &\quad = T_i^n + r\Theta_i^{n+1} + sg_{i+1}T_0^{n+1},\\ &\quad i = 2, 3, \ldots, M-1, n = 0, 1, \ldots, N-1.\end{aligned}$$

The above equations are expressed in matrix form

$$\mathbf{A}T^{n+1} = T^n + r\Theta^{n+1} + sT_0^{n+1}\mathbf{G} + sg_0T_M^{n+1}\mathbf{e}_{M-1}. \tag{15}$$

Here $\mathbf{A} = (A_{ij})$ is the matrix of coefficients such that

$$A_{ij} = \begin{cases} 0, & \text{when } j > i+1, \\ 1 - sg_1, & \text{when } i = j, \\ -sg_{i-j+1}, & \text{otherwise}, \end{cases}$$

and

$$T^n = \left[T_1^n, T_2^n, \ldots, T_{M-1}^n\right]^T, \quad \Theta^n = \left[\Theta_1^n, \Theta_2^n, \ldots, \Theta_{M-1}^n\right]^T,$$
$$\mathbf{G} = [g_2, g_3, \ldots, g_M]^T, \quad \mathbf{e}_{M-1} = [0, 0, \ldots, 1]^T.$$

Remark 9 Since the source term Θ appears in nonlinear form, we can solve it by iteration methods. Instead, in the numerical simulation we approximate $\Theta(T_i^{n+1})$ by replacing it with the corresponding value at previous time step $\Theta(T_i^n)$, viz.

$$\mathbf{A}T^{n+1} = T^n + r\Theta^n + sT_0^{n+1}\mathbf{G} + sg_0T_M^{n+1}\mathbf{e}_{M-1},$$

which is called an implicit–explicit (IMEX) method. One can prove that the above scheme is unconditionally stable and has the convergence rate of $O(\tau + h)$ under some reasonable assumptions (Liu et al. 2005).

One can refer to Tadjeran et al. (2006) for improving the numerical accuracy, where a second-order accurate numerical approximation for a spatial fractional diffusion equation was proposed. The approach based on the classical Crank–Nicolson method combined with spatial extrapolation was used to obtain temporally and spatially second-order accurate numerical estimates. It was shown that the fractional Crank–Nicolson method based on the shifted *Grünwald* formula is unconditionally stable. One can also obtain higher accuracy by applying predictor–corrector schemes.

4.2 Parameters and Conditions in Numerical Process

The fractional thermal conductivity κ_γ of textiles will be approximated by the classical case $\kappa = \varepsilon\kappa_a + (1-\varepsilon)\kappa_f$ (Note that they have different dimensions). In the simulation, we set

Thermal conductivities: $\kappa_a = 0.025\,\mathrm{W}\cdot\mathrm{m}^{-1}\cdot K^{-1}$, $\kappa_f = 0.1\,\mathrm{W}\cdot\mathrm{m}^{-1}\cdot K^{-1}$,
Thickness: $L = 2.5\times10^{-3}$ m, Time: $t \in [0, 10\,\mathrm{h}]$,
Left boundary value condition: $T_0 = 500\,°\mathrm{C}$,
Right boundary value condition: $T_1 = 37\,°\mathrm{C}$,
Initial condition:

$$T_I(x) = -\frac{T_0 - T_1}{L^2}x^2 + T_0,$$

The physical parameters in equations are given as follows (Yu and Xu 2015):

$$\varepsilon = 0.084,\ \tau = 1.2,\ \beta = 8\,\mathrm{m}^{-1},$$

$$C_v = 1715.0\mathrm{kJ} \cdot \mathrm{m}^{-3} \cdot \mathrm{K}^{-1},$$

$$\sigma = 5.672 \times 10^{-8}\mathrm{kJ} \cdot \mathrm{m}^{-2} \cdot \mathrm{K}^{-4} \cdot \mathrm{s}^{-1},$$

$$\xi_1 = \xi_2 = 0.9.$$

4.3 Numerical Result

In this subsection, the example shows that in high-temperature situation, the fractional model gives the more realistic results than the classical heat equation. We set $T_0 = 500\,^\circ\mathrm{C}$ and $T_1 = 37\,^\circ\mathrm{C}$. As was depicted in the introduction, the higher heat and humidity force us to use the superdiffusion model to simulate the heat transfer process for firefighter protective clothing during flash fire exposure. In the following, we will find that the superdiffusion model for this case conforms to the real situation. In Fig. 3a, the classical order $\gamma = 2.0$ is utilized, while the image with fractional order $\gamma = 1.8$ is shown on the right. It is obvious that the second case shows faster diffusion than the classical case. To clarify this result, we plot several curves in Fig. 4 with $\gamma = 2.0, 1.9, 1.8, 1.7, 1.6, 1.5$, respectively. As expected, the temperature $T(x, t)$ for fixed $t = t_{10} (M = 50, N = 50)$ drops faster and faster as γ is decreased when $\gamma > 1.5$, which is also valid for fixed $x = x_{10}$. It seems that $\gamma = 1.5$ is a critical value because the downward trend is no longer maintained when $\gamma \le 1.5$. Indeed, we observe that the trend is almost reversed in this case from Fig. 5. The above results indicate that it is appropriate to choose the fractional order γ such that $\gamma > 1.5$, which remains to be studied.

Remark 10 The superdiffusion model for the high-temperature case gives more reasonable results. On the other hand, the model describes the faster propagation as we expected.

We are focusing on the high temperature–humidity environment, where the firefighters put on protective clothing near the fire temperature higher than 500°C. It is hard to predict the lowest environmental temperature, before which the fractional model is no longer valid because of the complicated physical mechanism and different environmental parameters. Much luckily, in the case of example 2, we are sure to choose the lowest temperature $T_0 \ge 270\,^\circ\mathrm{C}$, here the fractional model is valid according to the numerical simulation.

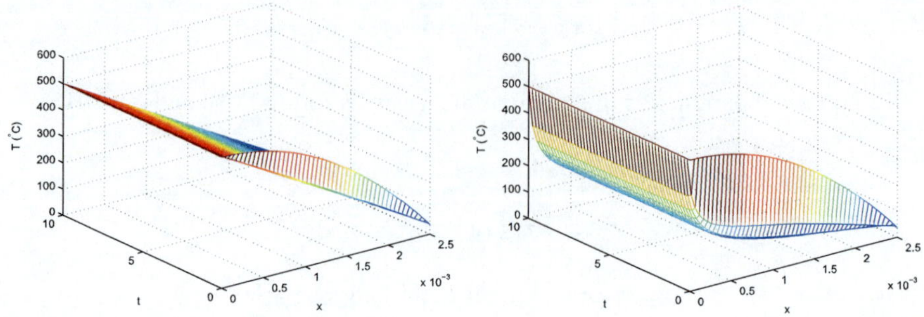

(a) $\gamma = 2.0$ (b) $\gamma = 1.8$

Fig. 3 $T(x, t)$ with $\gamma = 2.0$ and $\gamma = 1.8$ under high temperature

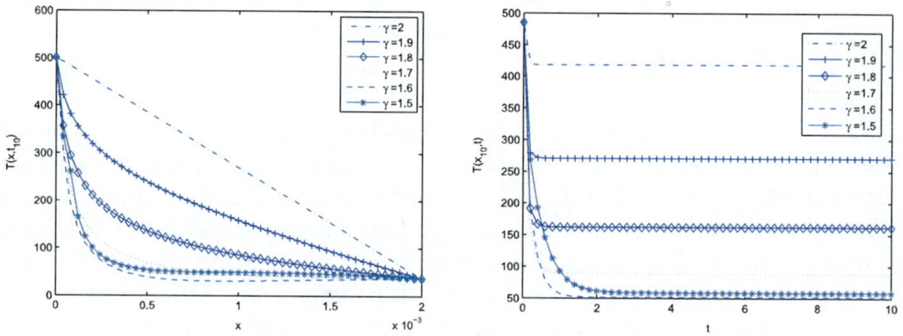

(a) $T(x, t_{10})$ (b) $T(x_{10}, t)$

Fig. 4 Temperature for fixed x and t when $\gamma \geq 1.5$

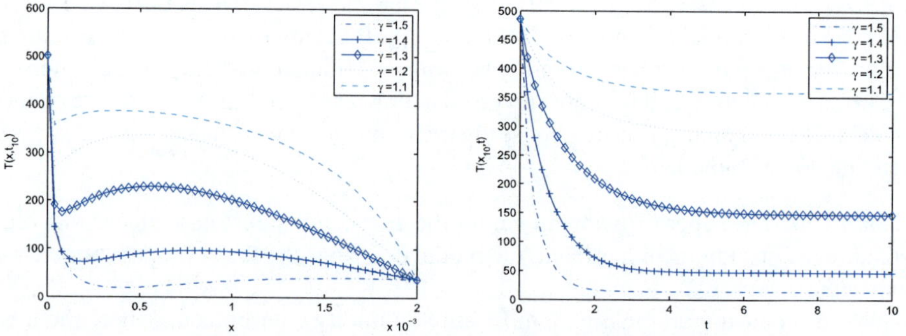

(a) $T(x, t_{10})$ (b) $T(x_{10}, t)$

Fig. 5 Temperature for fixed x and t when $\gamma \leq 1.5$

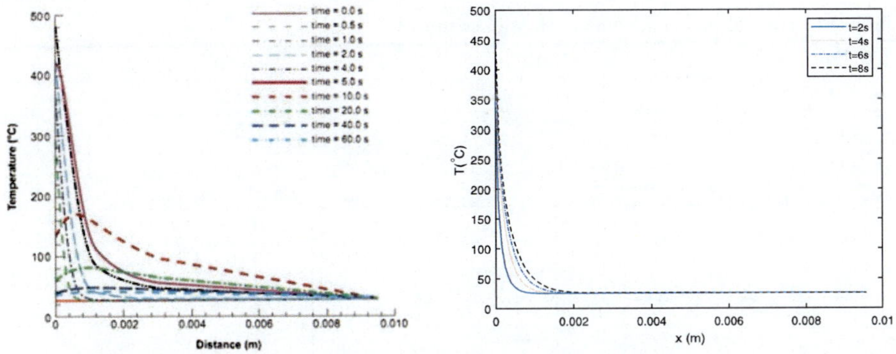

(a) Temperature measurements at sensors in [32]

(b) Temperature distribution by solving the DP 3

Fig. 6 Temperature distribution of measurement and numerical simulation

Remark 11 As depicted in Remark 5, Chitrphiromstri and Kuznetsov (2005) proposed a useful but complicated heat and moisture transfer model for firefighter protective clothing during a flash fire exposure. We find that the numerical images in Chitrphiromstri and Kuznetsov (2005) and Song et al. (2008) are qualitatively similar to our numerical results calculated by the simplified fractional model on the same conditions. See Fig. 6. Hence, it is expected to apply the fractional model to simplify the coupling phenomena with fewer parameters. To test the validity of the proposed fractional model or an improved fractional model which will be considered further, we shall carry out the experimental test on the numerical results observed in this paper.

In terms of skin burn evaluation, Henriques' integral model (Henriques and Moritz 1947) and Stoll's burn criterion (Stoll and Chianta 1969) can be used to predict the time to reach first-, second-, and third-degree burn injures. It is generally believed that the thermal damage of skin tissues occurs when the temperature of basal surface reaches 44^oC, and the degree of destruction continues to increase as the time prolongs (Chitrphiromstri and Kuznetsov 2005). Let $\Omega(L_{\text{bar}}, t)$ denote the value of Ω at the position of $x = L_{\text{bar}}$ (on the basal surface) during the time interval $[0, t_{\text{f}}]$. How to calculate the value of Ω?

Henriques' integral is introduced as follows

$$\Omega(x,t) = \int_0^t P\exp(-\frac{\Delta E}{R(T(x,\tau)+273.15)})\mathrm{d}\tau,$$

where Ω is the quantitative value of the skin burn, P is the frequency corruption factor, R is gas constant, and ΔE is the skin activation performance. All of their values can be found in Table 1.

Table 1 Values of parameters in the Henriques model

T/(°C)	P(1/s)	$\frac{\Delta E}{R}$/(K)
$44 \leq T < 50$	2.185×10^{124}	93261.9
$T \geq 50$	1.823×10^{51}	38836.8

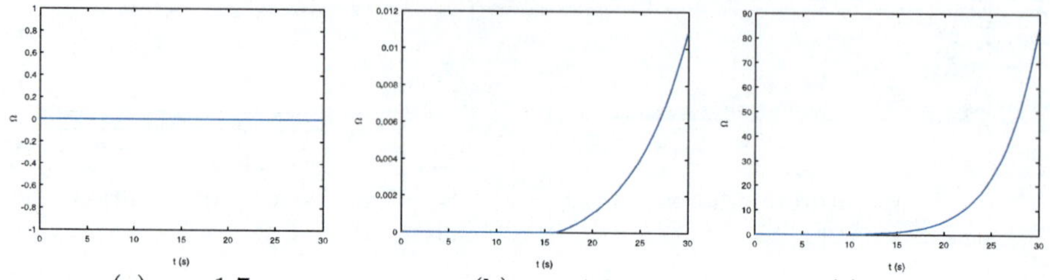

(a) $\gamma = 1.7$ (b) $\gamma = 1.8$ (c) $\gamma = 1.9$

Fig. 7 Value of $\Omega(L_{\text{bar}}, t)$ on the basel surface with different fraction indexes in the time interval [0, 30 s]

In Fig. 7, the value of $\Omega(L_{\text{bar}}, t)$ is becoming big as the time increases when $\gamma = 1.8,\ 1.9$, which indicates that the degree of burn injures is becoming serious. Due to the fact that the value of $\Omega(L_{\text{bar}}, t)$ depends on the temperature on the basel surface, the values of $\Omega(L_{\text{bar}}, t)$ with respect to the values of γ are consistent with the numerical results of temperature for fixed x in Fig. 4.

5 Mathematical Formulation for Inverse Problems for the TCC/TPC Design

5.1 *Mathematical Reformulation of the TCC Inverse Problems* (IP 1)

The mathematical formulation and classification of the IPTMD can be given according to the determination of single parameter and multiple parameters. For single-layered textile, the determined parameters include the thickness L, thermal conductivity k, and porosity ε of textiles. For multiple layered textiles, there are three parameters to be determined for each layer.

Define an operator equation

$$y = G_v(u) + \eta, \tag{16}$$

where $G_v : \mathbb{R}^3 \to \mathbb{R}^2$ is nonlinear mapping derived from the direct problem DP1 or DP2, $y = (T_b, RH_b)$, $v = (T_c, RH_c)$, η is the error of y; $u = (L, k, \varepsilon) \in \mathbb{R}^3$ is the

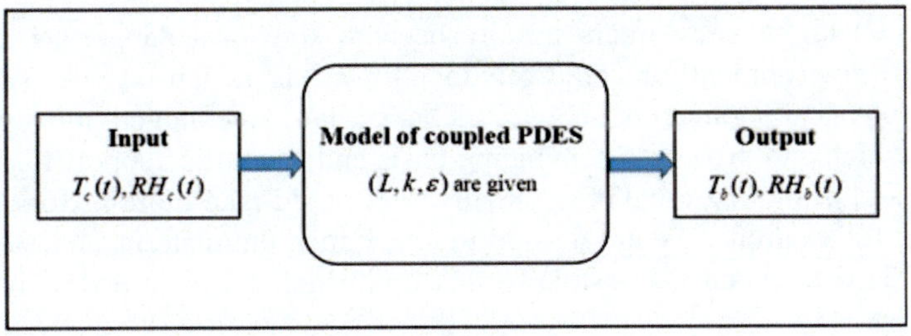

Fig. 8 Schematic diagram of the forward problems

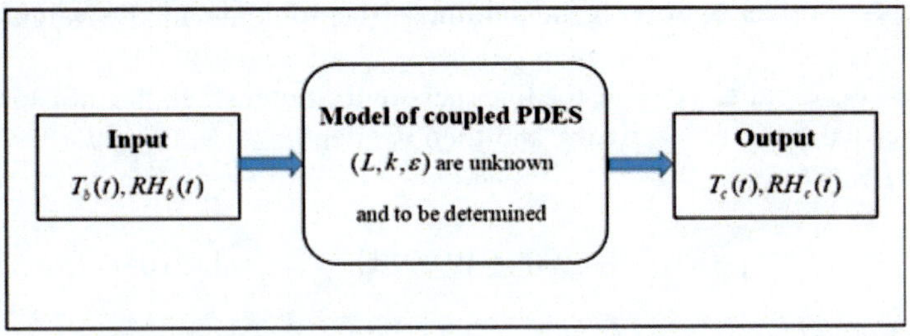

Fig. 9 Schematic diagram of the IPTMD

parameter vector which is related with the textiles and should be determined in the TCC design.

The forward problems can be described as follows: given $u = (L, k, \varepsilon)$ and G_v, η, we will determine y and make sure whether it is belong to the comfort index interval; see Fig. 8.

IPTMD: Given the environmental temperature and the relative humidity $(T_c(t), RH_c(t))$, we attend to determine the parameters of the TCC according to the thermal comfort indexes in the microclimate area, namely temperature (32 ± 1) °C, relative humidity $(50 \pm 10)\%$, and wind speed (25 ± 15) cm s^{-1}, respectively; see Fig. 9 for the schematic diagram of the IPTMD.

By means of solving **DP1/DP2** and **SP**, we should determine optimum $u = (L, k, \varepsilon) \in \mathbb{R}^3$ in the body–clothing–environment system to make sure that people feel thermally comfortable. That is to say, the values of temperature and humidity in the microclimate area are expected to lie stably in heat–moisture comfort indexes (intervals).

5.2 *Mathematical Reformulation of the TPC Inverse Problems* (IP 2)

How to design the TPC, which should protect fireworkers from suffering from thermal injures? It has been puzzling researchers and engineers.

As we all know, fireworkers usually need to stay in a dangerous and high-temperature environment for a relatively long time. It is of significance to determine optimally physical parameters, such as thickness, thermal conductivity, density, and porosity, which can provide a convincing basis and scientific support for choosing materials and designing the TPC. Actually, we should take more factors into consideration; for example, we are desired to economize the materials and reduce the weight of TPC as much as possible so that it is beneficial for fireworkers to move flexibly.

Aiming at the objectives of protecting fireworkers from thermal injuries, we determine triple thickness parameters simultaneously for the actual requirements of the TPC weights or prices by solving the minimization problem with the weighted objective function.

Let L_{out}, L_{bar}, and L_{lin} denote the thickness of the outer shell, the moisture barrier, and the thermal liner, respectively, and then we denote $L_1 = L_{\text{out}}$, $L_2 = L_1 + L_{\text{bar}}$ and $L_3 = L_2 + L_{\text{lin}}$.

Define

$$\Omega_0 = [0, 0.53), \quad \Omega_1 = [0.53, 1), \quad \Omega_2 = [1, 10^4),$$

which are critical intervals of no injury, the first-degree and the second-degree thermal injury on the basel surface, respectively. That is to say, no injury happens when $\Omega(L_{bas}, t_s)$ locates in $\Omega_0 = [0, 0.53)$ according to the experimental results; meanwhile first-degree thermal injury occurs when it locates in $\Omega_1 = [0.53, 1)$, and second-degree thermal injury occurs when it locates in $\Omega_2 = [1, 10^4)$ (Chitrphiromstri and Kuznetsov 2005).

Therefore, the inverse problem can be formulated to the following optimization problem.

$$\min \quad p_1 L_{\text{out}} + p_2 L_{\text{bar}} + p_3 L_{\text{lin}} \tag{17}$$

$$s.t. \ \ \Omega(L_5, t_s; L_{\text{out}}, L_{\text{bar}}, L_{\text{lin}}) \in \Omega_i, \tag{18}$$

where $i = 0, 1, 2$; p_1, p_2 and p_3 are the weights, which can be adjusted according to the actual requirements of fabric layers.

In details for better understanding, when it is required to ensure no thermal injury on the basal surface, (18) should be $\Omega(L_5, t_s; L_{\text{out}}, L_{\text{bar}}, L_{\text{lin}}) \in \Omega_0$. When the thermal injuries can be allowed to the first-degree but less than the second-degree injury, (18) should be $\Omega(L_5, t_s; L_{\text{out}}, L_{\text{bar}}, L_{\text{lin}}) \in \Omega_1$. When the thermal injuries can be allowed to the second-degree but less than the third-degree injury, (18) should be $\Omega(L_5, t_s; L_{\text{out}}, L_{\text{bar}}, L_{\text{lin}}) \in \Omega_2$.

6 Computational Strategy for the IPTMD of the TCC/TPC Design

6.1 *Deterministic Case: Least Squares Method and Regularization Method*

By the numerical solution to the **DP1/DP2** and **SP**, we can numerically obtain the temperature $T_0(t^j) = T(0, t^j)$ and humidity $C_0(t^j) = C_a(0, t^j)$ in the microclimate area $(x = 0)$ at any time t^j. Relative humidity $RH_0(t^j) = RH(0, t^j)$ is henceforth obtained according to Formula (6). $\mathrm{RH}_0(t^j)$ is rewritten as $\mathrm{RH}(0, t^j; u)$ since it varies as u. In general, we adopt relative humidity $\mathrm{RH}_A = 50\%$ in the sense of most comfortable expectation. The objective function of least squares is

$$J(u) = \sum_{j=1}^{n} \left|\mathrm{RH}(0, t^j; u) - \mathrm{RH}_A\right|^2.$$

In practical applications, there are some necessary requirements on textile material design. For example, we hope the clothing be kept much thinner and lower cost, so we need make limitation on the thickness of porous batting, i.e., the thickness satisfies $L \leqslant L_{\max}$ or $L_1 + L_2 \leqslant L_{\max}$. Similarly, we also hope the clothing be kept much lighter, that is to say, we have the limitation such as $\rho_1 L_1 + \rho_2 L_2 \leqslant K_{\max}$.

In order to solve the optimization problems of objective function with constraint conditions, we employ the regularization method, where we add a penalty term to the above objective function. Denote $J_\alpha(u)$ by

$$J_\alpha(u) = \sum_{j=1}^{n} \left|\mathrm{RH}(0, t^j; u) - \mathrm{RH}_A\right|^2 + \alpha \parallel u \parallel_*^2, \tag{19}$$

where $\parallel u \parallel_*$ is defined according to the practical situation of the vector $u = (L, k, \varepsilon)$.

We call the minimizer $u^\dagger$ of the objective function the regularized solution of the IPTMD if $u^\dagger$ satisfies

$$J_\alpha(u^\dagger) = \min J_\alpha(u), \tag{20}$$

where the parameters α is called the regularization parameter. The regularized solution of the above objective function is exactly the optimal solution which meet both thermal comfort indexes and above limitation conditions with respect to u.

The minimization problem $J_\alpha(u^\dagger) = \min J_\alpha(u)$ can be solved by some direct search methods such as Hooke–Jeeves methods and Golden section method. The optimal choice of the regularization parameter α can be determined by a prior choice or posterior choice such as L-curve method (Xu et al. 2013).

Remark 12 The regularization method can be effectively applied to the TPC design problems; for example, we can refer to Pan et al. (2017), Jiang et al. (2017)

6.2 Stochastic Case: Bayesian Inference Method and Maximum Probability Method

When η is random error, the above model

$$y = G_v(u) + \eta$$

is stochastic model, where $G : \mathbb{R}^3 \to \mathbb{R}^2$ is a nonlinear mapping, $u \in \mathbb{R}^3$ ia a random vector with probability density $\rho_0(u)$; η is a random noise with probability density $\rho(\eta)$, which is independent of u; y is the observational data. We will statistically deduce u from the given G_v, η and y.

Bayesian inference method (Xu et al. 2014):

Step 1 Give the prior density $\rho_0(u)$ of u.

Step 2 Compute the conditional probability density (likelihood function) $\rho(y - G_v(u))$ of $y \mid u$.

Step 3 Deduce the posterior density of u by Bayes' law

$$\rho^y(u) = \frac{\rho(y - G_v(u))\rho_0(u)}{\int\limits_{R^3} \rho(y - G_v(u))\rho_0(u)du}. \tag{21}$$

Step 4 Compute the point estimate or interval estimate of u.

Denote the potential function by

$$\Phi(u; y) = -\log \rho(y - G_v(u)).$$

Let μ^y be measure on $\mathbb{R}^3$ with density ρ^y, μ_0 be measure $\mathbb{R}^3$ with density ρ_0, then Bayes' theorem may be written as

$$\frac{\mathrm{d}\mu^y}{\mathrm{d}\mu_0}(u) = \frac{\mathrm{e}^{-\Phi(u;y)}}{\int\limits_{R^3} \mathrm{d}^{-\Phi(u;y)}\mu_0(\mathrm{d}u)}.$$

Consequently, the posterior measure μ^y is absolutely continuous in prior measure μ_0, the Radon–Nikodym derivative is proportional to likelihood function. Moreover, we have

$$E^{\mu^y} G_v(u) = E^{\mu_0}\left(\frac{\mathrm{d}\mu^y}{\mathrm{d}\mu_0}(u)G_v(u)\right).$$

Maximum probability method (Xu and Wen 2014):

Suppose $P(u)$ be the probability of an event that $\mathrm{RH}_0(t_j) \in [40\%, 60\%]$, for all j, when the vector u of fabric is preestablished. We want to find the optimal u, which has the maximum probability to make the human body feels comfortable. Hence, we consider the following maximization problem:

$$\text{maximize} \quad P(u). \tag{22}$$

Since the objective function $P(u)$ is not continuous, its points achieve maximum value may not be unique. Let $0 < M \leqslant 1$ be the maximum value of $P(u)$. Considering that the smaller value of $\| u \|_*$ the better the human body feels. Therefore, the original optimization problem (22) can be modified as

$$\text{minimize} \quad \| u \|_*, \quad \text{s.t.} \quad P(u) = M. \tag{23}$$

This objective function is linear and combined with the nonlinear constraint. Suppose u be the solution to this constrained optimization problem, then we call it a maximum probability solution with minimum norm.

Using static penalty method, the above-constrained problem can be replaced by the following unconstrained problem

$$\text{minimize} \quad \| u \|_* + \text{K} \times [M - P(u)], \tag{24}$$

where K is a large positive constant.

Since the objective function of the optimization problem (24) is not continuous, conventional optimization technique such as gradient-based algorithm is not good enough to solve the problems involving the probability function. We can employ a new stochastic method known as particle swarm optimization(in abbreviation PSO) algorithm to solve the above optimization problem.

The advantages of the PSO algorithm are its capability in searching for the global optimum and no computation of the complicated gradients. Because of its simplicity of implementation as well as ability to swiftly converge to a good solution, the algorithm only requires fitness function for each of the particle, without assumption such as continuity and differentiability, which makes it very useful for a discontinuous function.

Remark 13 The stochastic methods can be effectively applied to the TCC design problems; for example, we can refer to the reference Xu et al. (2018). Numerical simulation results can be found in Xu et al. (2013), Xu et al. (2014), Xu and Wen (2014), Xu (2014), Xu et al. (2015), Yu and Xu (2015), Xu and Cui (2016), Ge et al. (2017). In the near future, the stochastic methods will be employed to solve the TPC design problems.

6.3 Computational Examples for IPTMD

In this subsection, we implement computational examples of single parameter determination and multiple parameters determination for (**IP 1**).

Figure 10 shows the graph of $J(L)$ with various thicknesses of the fiber, and it concludes that we can search the unique minimum point of $J(L)$ by the dichotomy of

Fig. 10 J(L)

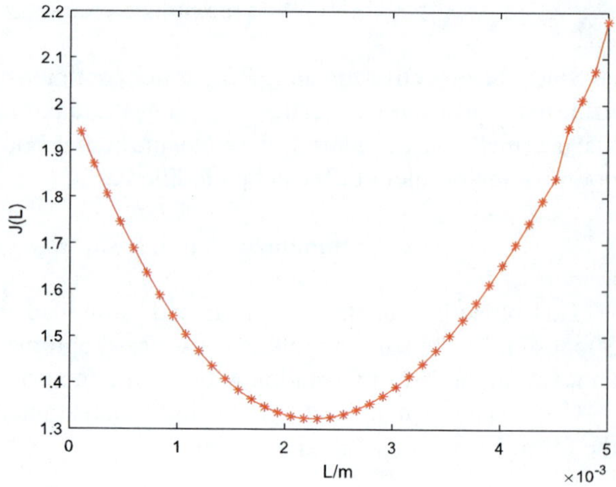

the algorithm effectively. Satisfying the accuracy requirement 1×10^{-4}, the approximate optimal thickness of the fabric is 2.3413mm.

Figure 11a–c shows the graphs of $J(k, L)$, $J(k, \varepsilon)$, $J(\varepsilon, L)$ with various parameters, respectively. It concludes that it is likely that the heat conductivity k makes less impact on the value of J than the thickness L and the porosity ε. Therefore, the problem of the thickness and the heat conductivity determination can be simplified to the problem of the single parameter determination. Besides, the problem of the porosity and the heat conductivity determination is similar. Next, we can search the unique minimum point of $J(\varepsilon, L)$ by implementing the appropriate algorithm effectively and obtain the approximate optimal combination of porosity and thickness for the fabric (0.17, 5 mm).

7 Concluding Remarks and Future Studies

The well-posedness results of the **DP1**, **DP2**, and **DP3** verify the existence, uniqueness, and stability of solution to the heat–moisture transfer model under suitable assumptions, which provides a theoretical foundation to the numerical solutions of the **DP1**, **DP2**, and **DP3**. Numerical simulation is to achieve the distribution of temperature and water vapor concentration in porous batting, which helps us to judge whether the body heat–moisture comfort in the microclimate area arrives. This belongs to the direct problem approaches.

In the view of the inverse problem approaches, i.e., IPTMD, the optimal determination of the textile parameters u can be reformulated as inverse problems **IP1** and **IP2**, and their numerical solution can achieve the goal of parameter determination. The numerical results prove the feasibility of the theory and algorithm of the IPTMD. Moreover, the numeral results confirm rationality of the objective function

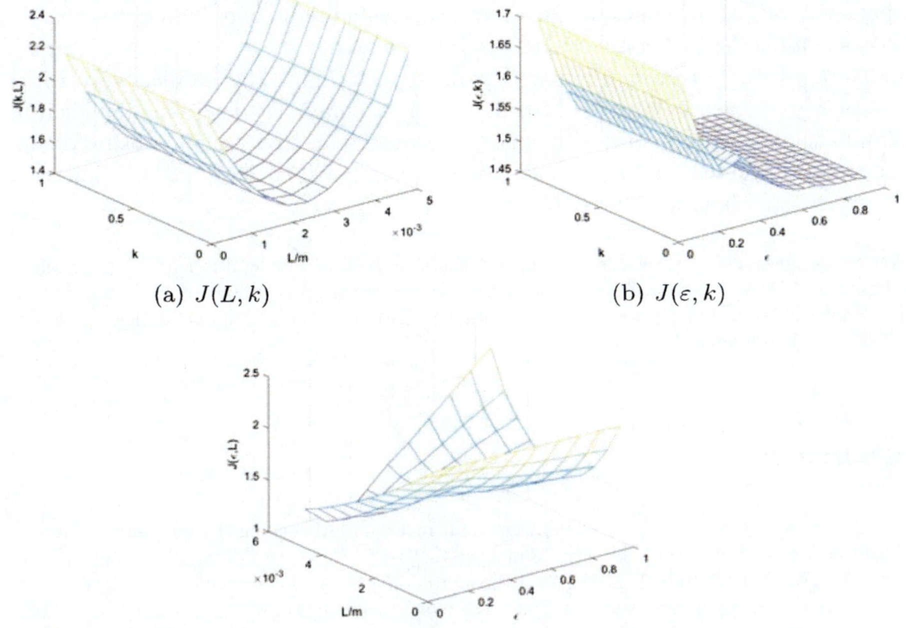

(a) $J(L, k)$ (b) $J(\varepsilon, k)$

(c) $J(\varepsilon, L)$

Fig. 11 Temperature distribution of measurement and numerical simulation

and stability of the algorithm of the IPTMD, which provides theoretical basis to new TCC/TPC material development. For the future researches, we should continue to discuss the well-posedness of the inverse problems **IP1** and **IP2**.

For the coupled nonlinear partial differential equations based on the dynamic heat–moisture transfer process, we will continue to develop more efficient numerical algorithms for the TCC development. For fractional heat transfer models aiming to simulate the situation with high temperature and high humidity, we haven't found a similar related fractional model in heat and moisture transfer within textiles, which may be result from the fact that the simulation results go against the real situation under low temperature. Therefore, we will focus on discussing the phase change, due to sorption and condensation, in the TPC. We are also interested in the inverse problems of heat and moisture transfer model for TPC system. In this new context, the formulation of inverse problems will be very different from the classical model considered in Xu (2014), Xu et al. (2010), Xu and Ge (2012), Xu et al. (2014), Xu and Wen (2014) because sweat cannot be neglected and we are more concerned with thermal damage in this case. The regularity of the weak solution should be considered. One can refer to Jin et al. (2015) for additional information. The finite element approximation for the variational formulation reported in the paper also deserves further consideration.

We are urgent to derive the statistical inversion of the IPTMD due to the stochastic characteristics of the environmental temperature–humidity and the interval characteristics of the thermal comfort/safety index, and realize the multi-parameters determination of multilayer textile materials for the TCC and TPC, respectively. More practically, we are willing to solve the multiple parameter determination such as simultaneous determination of thickness, thermal conductivity, and porosity for single or multilayer textile materials. Practical software will be developed to implement numerical simulations of the IPTMD.

Acknowledgements This research is supported by the National Natural Science Foundation of China (No. 11871435 and 11471287). The work was elected as a plenary talk on the Forum "Math-for-Industry" 2018– Big Data Analysis, AI, Fintech, Math in Finances and Economics, November 17–21, 2018, Shanghai, China.

References

Bowles M, Agueh M (2015) Weak solutions to a fractional Fokker-Planck equation via splitting and Wasserstein gradient flow. Appl Math Lett 42:30–35

Chen YB, Xu DH, Zhou XH (2011) An inverse problem of type determination for bilayer textile materials under low temperature. In: Symposium on Proceedings of bioengineering and information society, Text, pp 1336–1343

Chitrphiromstri P, Kuznetsov AV (2005) Modeling heat and moisture transfer in firefighter protective clothing during flash fire exposure. Heat Mass Transfer 41:206–215

Du N, Fan JT, Wu HJ, Sun WW (2009) Optimal porosity distribution of fibrous insulation. Int J Heat Mass Trans 52:4350–4357

Du Q, Gunzburger M, Lehoucq RB, Zhou K (2012) Analysis and approximation of nonlocal diffusion problems with volume constraints. SIAM Rev 54:667–696

Elgafy A, Mishra S (2014) A heat transfer model for incorporating carbon foam fabrics in firefighter's garment. Heat Mass Transfer 50:545–557

Engl HW, Hanke M, Neubaer A (1998) Regularization of inverse problems. Kluwer Academic Publishers, London

Ervin V, Roop J (2006) Variational formulation for the stationary fractional advection dispersion equation. Numer Meth Part D E 22:558–576

Fan JT, Wei XH (2002) Heat and moisture transfer through fibrous insulation with phase change and mobile condensates. Int J Heat Mass Transfer 19:4045–4055

Fan JT, Luo ZX, Li Y (2000) Heat and moisture transfer with sorption and condensation in porous clothing assemblies and numerical simulation. Int J Heat Mass Transfer 43:2989–3000

Fan JT, Cheng X, Wen X, Sun WW (2004) An improved model of heat and moisture transfer with phase change and mobile condensates in fibrous insulation and comparison with experimental results. Int J Heat Mass Transfer 47:2343–2352

Friedman A (1964) Partial differential equations of parabolic type. Prentice-Hall Inc

Ge MB, Yu Y, Xu DH (2017) Textile porosity determination based on a nonlinear heat and moisture transfer model. Appl Anal 96:1681–1697

Henriques FC, Moritz AR (1947) Studies of thermal injuries I: the conduction of heat to and through skin and the temperatures attained therein. Theor Exp Invest Am J Pathol 23:531–549

Hua YX, Yu XH (2013) On the ground state solution for a critical fractional Laplacian equation. Nonl Anal 87:116–125

Huang J (2008) Cloting comfort. Scienc Press, Beijing

Jiang XY, Xu DH, Zhang QF (2017) A modified regularized algorithm for a semilinear space-fractional backward diffusion problem. Math Method Appl Sci 40:5996–6006
Jin BT, Lazarov R, Pasciak J, Rundell W (2015) Variational formulation of problems involving fractional order differential operators. Math Comput 84:2665–2700
Klafter J, Sokolov IM (2005) Anomalous diffusion spreads its wings. Phys World 8:29–32
Li Y, Li FZ, Liu YX, Luo ZX (2004) An integrated model for simulating interactive thermal processes in human-clothing system. J Therm Biol 29:567–575
Liu F, Zhuang P, Anh V, Tuner I (2005) A fractional-order implicit difference approximation for the space-time fractional diffusion equation. Anziam J 47:C48–C68
Metzler R, Klafter J (2000) The random walk's guide to anomalous diffusion: a fractional dynamics approach. Phys Rep 339:1–77
Metzler R, Klafter J (2004) The restaurant at the end of the random walk: recent developments in the description of anomalous transport by fractional dynamics. J Phys A: Math Gen 37:R161–R208
Metzler R, Klafter J, Sokolov IM (1998) Anomalous transport in external fields: Continuous time random walks and fractional diffusion equations extended. Phys Rev E 58:1621–1632
Pan B, Xu DH, Xu YH, Yu Y (2017) TV-like regularization for backward parabolic problems. Math Meth Appl Sci 40:957–969
Podlubny I (1999) Fract differential equations. Academic Press, New York
Song GW, Chitrphiromstri P, Ding D (2008) Numerical simulations of heat and moisture transport in thermal protective clothing under flash fire conditions. Int J Occup Saf Ergo 14:89–106
Stoll AM, Chianta MA (1969) Method and rating system for evaluation of thermal protection. Aerosp Med 11:1232–1238
Stuart AM (2010) Inverse problems: a Beyesian perspective. Acta Numer 19:451–559
Tadjeran C, Meerschaert MM, Scheffler HP (2006) A second-order accurate numerical approximation for the fractional diffusion equation. J Comput Phys 213:205–213
Wu HH, Fan JT (2008) Study of heat and moisture transfer within multi-layer clothing assemblies consisting different types of battings. J Therm Sci 47:641–647
Xu DH, Cheng JX, Chen YB, Ge MB (2011) An inverse problem of thickness design for bilayer textile materials under low temperature. J Phys: Conf Ser 290:12018
Xu DH, Cui P (2016) Simultaneous determination of thickness, thermal conductivity and porosity in textile material design J Inverse Ill-pose 24:59–66
Xu DH, Wen L (2014) An inverse problem of textile porosity determination Chin Ann Math 35(A):129–144
Xu DH (2014) Mathemtical modeling of heat and moisture transfer within textiles and corresponding inverse problems of textile material design. Science Press, Beijing
Xu DH (2014) Inverse problem of textile material design based on clothing heat-moisture comfort. Appl Anal 93:2426–2439
Xu DH, Ge MB (2012) Thickness determination in textile material design: dynamic modeling and numerial algorithms. Inverse Probl 28:35011–35032
Xu DH, Chen YB, Zhou XH (2010) An inverse problem of thickness design for single layer textile material under low temperature. J Math Ind 2:582–590
Xu DH, Chen RL, Ge MB (2012) Inverse problems of textile material design based on comfort of clothing. Commun Appl Comput Math 3:332–341
Xu DH, Chen YB, Zhou XH (2013) Type design for textile materials under low temperature: modeling, numerical algorithm and simulation. Int J Heat Mass Transfer 60:582–590
Xu DH, Wen L, Xu BX (2014) An inverse problem of bilayer textile thickness determination in dynamic heat and moisture transfer. Appl Anal 93:445–465
Xu YH, Xu DH, Zhang LP, Zhou XH (2015) A new inverse problem for the determination of textile fabrics. Inverse Probl Sci Eng 23:635–650
Xu DH, He YG, Yu Y, Zhang QF (2018) Multiple parameter determination in textile material design: a Bayesian inference approach based on simulation. Math Comput Simul 151:1–14
Yang GF, Yamamoto M, Cheng J (2008) Heat transfer in composite materials with Stefan-Boltzmann interface conditions. Math Meth Appl Sci 31:1297–1314

Ye C, Huang HX, Fan JT, Sun WW (2008) Numerical study of heat and moisture transfer in textile materials by a finite volume method. Commun Comput Phys 4:929–948

Yosida K (1999) Functional analysis (sixth version). Springer, Beijing World Publishing Corporation, Beijing

Yu Y, Xu DH (2015) On the inverse problems of thermal conductivity determination in nonlinear heat and moisture transfer model within textiles. Appl Math Comput 264:284–299

Yu Y, Xu DH, Hon YC (2015) Reconstruction of inaccessible boundary value in a sideways parabolic problem with variable coefficients. Eng Anal Bound Elem 61:78–90

Yu Y, Xu DH, Hon YC (2015) Numerical algorithms for a sideways parabolic problem with variable coefficients. Appl Anal 95:874–901

Yu Y, Xu DH, Steve Xu YZ, Zhang QF (2016) Variational formulation for a fractional heat transfer model in firefighter protective clothing. Appl Math Model. 40:9675–9691

Determinantal Reinforcement Learning with Techniques to Avoid Poor Local Optima

Takayuki Osogami and Rudy Raymond

1 Introduction

Reinforcement learning for multiple collaborative agents is important and has many practical applications. Consider a team that consists of different types of players. In many cases, to win a game, each type of player must master highly relevant actions that are necessarily different from the other types of player. For example, players of a defensive team should guard relevant and diverse areas or relevant and different types of players of the other team. This is also the case when controlling many similar robots to perform a task that cannot be performed by a single one. Training them in a way they take different and relevant actions can lead to faster and better convergence to optimal collaboration.

Typical approaches in reinforcement learning, which let each agent take actions independently of other agents, are therefore not effective. Formulating the learning as handling the combination of actions as if it is an action of a hypothetical agent can lead to searching in an exponentially larger action spaces that grow with the number of agents and therefore does not scale well.

In Osogami and Raymond (2019), we have proposed the use of the determinant of a matrix to approximate the action-value function in reinforcement learning that takes into account both relevance and diversity in a natural manner. When each action of an agent in a particular state is characterized by a feature vector, the length of the feature vector corresponds to the relevance of that action at that state. Meanwhile, the angle between two feature vectors represents the diversity between the two corresponding actions at that state. A set of feature vectors then comprises a parallelotope whose volume is determined by the lengths (i.e., relevances) and the angles (i.e., diversities) of

T. Osogami (✉) · R. Raymond
IBM Research—Tokyo, Tokyo, Japan
e-mail: osogami@jp.ibm.com

R. Raymond
e-mail: rudyhar@jp.ibm.com

J. Cheng et al. (eds.), *Proceedings of the Forum "Math-for-Industry" 2018*,
Mathematics for Industry 35, https://doi.org/10.1007/978-981-16-5576-0_5

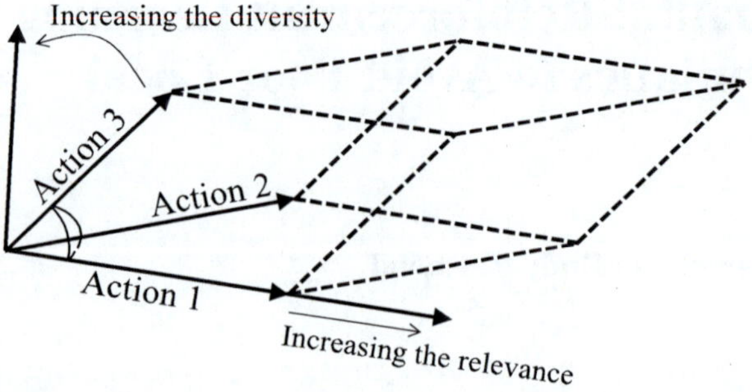

Fig. 1 The logarithm of the squared volume of the parallelepiped defined by the feature vectors of actions represents the value of the combination of those actions. The volume of the parallelepiped is determined by the length of vectors (their relevances) and their angles (their diversities). The volume can be increased by increasing relevance, diversity, or both

the feature vectors. An example of a parallelotope from three feature vectors in three dimensions is shown in Fig. 1. The figure also illustrates two ways the volume can be increased: increasing relevance and increasing diversity. The squared volume of the parallelotope can be computed from the determinant of the Gram matrix of the feature vectors. More specifically, the value of a combination of relevant and diverse actions at a state can be computed from the logarithm of the determinant (log-determinant) of the principal submatrix of a positive semidefinite matrix (kernel), where the principal submatrix is specified by the actions, and the kernel depends on the state.

In Osogami and Raymond (2019), we have derived efficient learning rules of determinantal SARSA (state-action-reward-state-action algorithm). Namely, we approximate the action-value function of N agents with an $N \times N$ kernel matrix whose effective dimension is $K \ll N$, so that at each iteration the action-value function can be updated with the additional $O(K^3)$ computational complexity. Determinantal SARSA has been shown to find nearly optimal policies approximately ten times faster than baseline approaches (Sallans and Hinton 2001, 2004; Heess et al. 2017; Sallans 2002), where free energy of a restricted Boltzmann machine (RBM) is used as a functional approximator. In this paper, we add theoretical insights and experiments on how to avoid policies with poor local optima by techniques based on derivation of the learning rules of determinantal SARSA.

2 Determinantal SARSA

In this section, we review determinantal SARSA, which we have proposed in Osogami and Raymond (2019). Determinantal SARSA considers the setting with a team of agents under central control. In the following, an agent team refers to the team of agents, and a team action refers to the combination of their actions. At each time t, the agent team makes an observation o_t. Let $\mathbf{z}_t \equiv \xi(a_{t-1}, r_t, o_t)$ represent the (features of) observation at time t, which may include the preceding team action a_{t-1}

and reward r_t in addition to o_t. Let $\mathbf{z}_{\leq t}$ denote the observations up to t. Depending on what has been observed (i.e., $\mathbf{z}_{\leq t}$), the agent team takes a team action a_t. Let $\mathbf{x}_t \equiv \psi(a_t) \in \{0, 1\}^N$ be a binary representation of a team action a_t (e.g.,, $\mathbf{x}_t$ may indicate which subset of N possible actions is taken by the agent team). The agent team then obtains reward r_{t+1}, and the environment changes its state. The agent team then makes a partial observation o_{t+1} of the environment and chooses the next team action a_{t+1}, and this process is continued. The goal of the agent team is to sequentially choose team actions so that the cumulative reward is maximized.

Given that the agent team has $\mathbf{z}_{\leq t}$, performs the action with the binary representation $\mathbf{x}$, and acts according to the policy under consideration, determinantal SARSA seeks to learn the Q (action value) function $Q_\theta(\mathbf{z}_{\leq t}, \mathbf{x})$ so that it best approximates the expected cumulative reward. By learning the Q function, one can identify the action that is optimal at a given state when one follows the policy under consideration from the next state. This allows one to iteratively improve the policy under consideration.

In determinantal SARSA, the Q function is assumed to have the following form:

$$Q_\theta(\mathbf{z}_{\leq t}, \mathbf{x}_t) \equiv \alpha + \log\det \mathbf{V}(\mathbf{x}_t)\,\mathrm{Diag}(\exp(\mathbf{d}_t(\phi)))\,\mathbf{V}(\mathbf{x}_t)^\top. \tag{1}$$

Here, $\mathbf{V}$ is an arbitrary $N \times K$ matrix for $0 < K \leq N$, and $\mathbf{V}(\mathbf{x}_t)$ denotes the matrix consisting of a subset of the rows of $\mathbf{V}$ in a way that the rows of $\mathbf{V}(\mathbf{x}_t)$ are indexed by the elements that are one in $\mathbf{x}_t$. Also, $\mathrm{Diag}(\cdot)$ denotes the diagonal matrix formed with a given vector, $\mathbf{d}_t(\phi)$ is a time-varying K-dimensional vector, and its exponentiation is elementwise. Here, $\mathbf{d}_t(\phi)$ should be considered as a time-series model, with parameter ϕ, that outputs a K-dimensional vector. Also, $\mathbf{d}_t(\phi)$ should be differentiable with respect to ϕ to allow end-to-end learning. Examples of such $\mathbf{d}_t(\phi)$ include recurrent neural networks (Hausknecht and Stone 2015), vector autoregressive models, and dynamic Boltzmann machines (Osogami and Otsuka 2015; Osogami 2017).

To intuitively understand the form of Q_θ in (1), consider the case where $\mathbf{V}$ is the identity matrix of order $K = N$. In this case, Q_θ is reduced to

$$Q_\theta(\mathbf{z}_{\leq t}, \mathbf{x}_t) = \alpha + \mathbf{d}_t(\phi)^\top \mathbf{x}. \tag{2}$$

If the i-th element of $\mathbf{x}$ indicates whether the i-th action is taken by an agent, the value of a team action is the sum of the values of individual actions without consideration of diversity, where $\mathbf{d}_t(\phi)$ represents the value (relevance) of individual actions at time t. With a non-identity $\mathbf{V}$, determinantal SARSA can take into account the diversity in actions.

Determinantal SARSA learns all of the parameters $\theta \equiv (\alpha, \mathbf{V}, \phi)$ in an end-to-end manner. Specifically, at each iteration, determinantal SARSA updates θ according to

$$\theta \leftarrow \theta + \eta\,\big(r_{t+1} + \rho\, Q_\theta(\mathbf{z}_{\leq t+1}, \mathbf{x}_{t+1}) - Q_\theta(\mathbf{z}_{\leq t}, \mathbf{x}_t)\big)\,\nabla_\theta Q_\theta(\mathbf{z}_{\leq t}, \mathbf{x}_t), \tag{3}$$

where we need the gradient $\nabla_\theta Q_\theta$. In Osogami and Raymond (2019), we have shown that the gradient can be represented in a computationally convenient form as follows:

Algorithm 1 Determinantal SARSA (Osogami and Raymond 2019)[1]

1: **Input:** Discount factor ρ; learning rate η; initial θ
2: Take initial team-action a_0; $\mathbf{x}_0 \leftarrow \psi(a_0)$
3: **for** $t = 0, 1, \ldots$ **do**
4: Get r_{t+1} and observe o_{t+1}; $\mathbf{z}_{t+1} \leftarrow \xi(a_t, r_{t+1}, o_{t+1})$
5: Take team-action a_{t+1}; $\mathbf{x}_{t+1} \leftarrow \psi(a_{t+1})$
6: $\mathbf{D}_t \leftarrow \mathrm{Diag}(\exp(\mathbf{d}_t(\phi)))$
7: Update $\mathbf{d}_t(\phi)$ to $\mathbf{d}_{t+1}(\phi)$ with $\mathbf{z}_{t+1}$
8: $\mathbf{D}_{t+1} \leftarrow \mathrm{Diag}(\exp(\mathbf{d}_{t+1}(\phi)))$
9: $Q_t \leftarrow \alpha + \log\det \mathbf{V}(\mathbf{x}_t)\, \mathbf{D}_t\, \mathbf{V}(\mathbf{x}_t)^\top$
10: $Q_{t+1} \leftarrow \alpha + \log\det \mathbf{V}(\mathbf{x}_{t+1})\, \mathbf{D}_{t+1}\, \mathbf{V}(\mathbf{x}_{t+1})^\top$
11: $\Delta_t \leftarrow r_{t+1} + \rho\, Q_{t+1} - Q_t$
12: $\alpha \leftarrow \alpha + \eta\, \Delta_t$
13: $\mathbf{V}(\mathbf{x}_t) \leftarrow \mathbf{V}(\mathbf{x}_t) + 2\,\eta\, \Delta_t\, (\mathbf{V}(\mathbf{x}_t)^+)^\top$
14: $\phi \leftarrow \phi + \eta\, \Delta_t\, \mathrm{diag}\left(\mathbf{V}(\mathbf{x}_t)^+\, \mathbf{V}(\mathbf{x}_t)\right)\, \nabla_\phi \mathbf{d}_t(\phi)$
15: **end for**

$$\nabla_\alpha Q_\theta(\mathbf{z}_{\le t}, \mathbf{x}) = 1 \tag{4}$$

$$\nabla_{\mathbf{V}(\bar{\mathbf{x}})} Q_\theta(\mathbf{z}_{\le t}, \mathbf{x}) = \mathbf{0} \tag{5}$$

$$\nabla_{\mathbf{V}(\mathbf{x})} Q_\theta(\mathbf{z}_{\le t}, \mathbf{x}) = 2\,(\mathbf{V}(\mathbf{x})^+)^\top \tag{6}$$

$$\nabla_\phi Q_\theta(\mathbf{z}_{\le t}, \mathbf{x}) = \mathrm{diag}\left(\mathbf{V}(\mathbf{x})^+\, \mathbf{V}(\mathbf{x})\right)\, \nabla_\phi \mathbf{d}_t(\phi) \tag{7}$$

where $\mathbf{V}(\mathbf{x})^+$ denotes the pseudo-inverse of $\mathbf{V}(\mathbf{x})$, $\mathrm{diag}(\cdot)$ denotes the vector formed with the diagonal elements of a given matrix, and $\bar{\mathbf{x}} \equiv 1 - \mathbf{x}$ elementwise.

Algorithm 1 gives a pseudocode of determinantal SARSA. In each iteration of the for-loop starting at Step 3, after getting reward r_{t+1} and making an observation o_{t+1} in Step 4, one takes a team action a_{t+1} in Step 5. Steps 6–8 compute the diagonal matrix $\mathbf{D}_t \equiv \mathrm{Diag}(\exp(\mathbf{d}_t(\phi)))$ by using the time-series model $\mathbf{d}_t(\phi)$, whose state is updated in Step 7 on the basis of the input $\mathbf{z}_t$. These diagonal matrices are then used in Steps 9–12 to compute the TD error Δ_t. The parameters $\theta \equiv (\alpha, \mathbf{V}, \phi)$ are then updated in Steps 12–14. In Step 14, the gradient $\nabla_\phi \mathbf{d}_t(\phi)$ depends on the particular time-series model under consideration. It is easy to estimate the computational time to update parameters at each iteration of determinantal SARSA. Since the rank of $\mathbf{V}$ is at most K, the computational complexity of the pseudo-inverse $\mathbf{V}(\mathbf{x}_t)^+$ and the computation of $\log\det \mathbf{V}(\mathbf{x}_t)\mathbf{D}_t\mathbf{V}(\mathbf{x}_t)^\top$ is $O(K^3)$.

Note that one may use determinantal SARSA to the fully observable case by letting $\mathbf{D}_t \equiv \mathrm{Diag}(\exp(\mathbf{d}_t(\phi)))$ in (1) be static but depend on the fully observed Markovian state s_t at time t. For example, one may use a feedforward neural network $\psi(\cdot)$ that maps a state s_t into a K-dimensional feature vector $\mathbf{d} = \psi(s_t)$, which then defines $\mathbf{D}_t = \mathrm{Diag}(\exp(\mathbf{d}))$. With $\mathbf{D}_t$ alone, the state can only influence the values

[1] Here, typographical errors in [10] are corrected by adding "⊤" in Step 9–10 and removing "¬" in Step 13. In Step 13–14, the $\mathbf{V}(\mathbf{x}_t)^+$ is the pseudo inverse that will be slightly modified for improving the stability of Determinantal SARSA.

of individual actions, while we also allow the state to affect the diversity measure as well through $\mathbf{V}$.

In Step 2 and Step 5 of Algorithm 1, we need to choose team actions in consideration of the tradeoff between exploration and exploitation. One of popular approaches is Boltzmann exploration. In Osogami and Raymond (2019), we have shown that, for determinantal SARSA, the Boltzmann exploration with unit temperature reduces to sampling from a determinantal point process, which allows efficient (in time polynomial in N) sampling (Kulesza and Taskar 2012; Qiao et al. 2016; Kulesza and Taskar 2011). The Boltzmann exploration with general temperature for determinantal SARSA requires sampling from annealed determinantal distributions (Wachinger and Golland 2015; Belabbas and Wolfe 2009), for which practical sampling algorithm as Markov Chain Monte Carlo is available (Kang 2013; Gillenwater 2014).

3 Avoiding Poor Local Optima

In the experiments of Osogami and Raymond (2019), we have occasionally observed that determinantal SARSA is trapped into poor local optima. We hypothesize that this can happen because of the low-rank kernel approximation and the pseudo-inverse updates. Here, we show how we can avoid such poor local optima.

Our formulation of (1) assumes that the rank of the following $N \times N$ positive semidefinite matrix (kernel)

$$\mathbf{L}_t \equiv \mathbf{V}\,\mathrm{Diag}(\exp(\mathbf{d}_t(\phi)))\,\mathbf{V}^\top \tag{8}$$

is K. This is achieved by the use of the $N \times K$ matrix $\mathbf{V}$. However, depending on the initial values of $\mathbf{V}$, determinantal SARSA may fall into the situation where the rank of $\mathbf{V}$ (and its submatrix $\mathbf{V}(\mathbf{x})$) becomes smaller than K. Once determinantal SARSA is trapped into such $\mathbf{V}$s, it cannot search for parameters on the larger subspace. Namely, with (6), the parameters $\mathbf{V}(\mathbf{x})$ are updated by adding the terms that are proportional to $\left(\mathbf{V}(\mathbf{x})^+\right)^\top$. However, because of the property of the pseudo-inverse, we can observe that

$$\mathrm{Range}\,(\mathbf{V}(\mathbf{x})) = \mathrm{Range}\left(\left(\mathbf{V}(\mathbf{x})^+\right)^\top\right), \tag{9}$$

where Range $(\mathbf{A})$ is the range or the image of a matrix $\mathbf{A}$. Thus, determinantal SARSA may not be able to reach optimal solutions from some initial values. This explains why determinantal SARSA can be trapped into local optima.

The above observation leads to mitigation techniques by keeping the rank of $\mathbf{V}(\mathbf{x})$ to be K during the computation. This can be achieved heuristically by adding some noises to $\mathbf{V}$ in the initialization and to the pseudo-inverse $\mathbf{V}(\mathbf{x})^+$ in each iteration.

First, we propose to initialize $\mathbf{V}$ using random special orthogonal matrices (Stewart 1980), $\mathbf{A}$ and $\mathbf{B}$, where $\mathbf{A}$ is $N \times N$, and $\mathbf{B}$ is $K \times K$. Specifically, we initialize $\mathbf{V}$ as

$$\mathbf{V} = \mathbf{A}\,\Sigma\,\mathbf{B}^\top + \mathcal{E}, \tag{10}$$

where Σ is an $N \times K$ rectangular diagonal matrix in which every diagonal element is one, $\mathcal{E}$ is a random $N \times K$ matrix (in our experiments, and we will sample each element independently according to the uniform distribution with support $[-0.01, 0.01]$). Namely, the matrix $\mathbf{V}$ is initialized in a way such that each of its singular values is one with small noise $\mathcal{E}$. This particular initialization plays a rather important role in avoiding convergence to poor local optima.

Next, let

$$\mathbf{V} = \mathbf{A}\,\Sigma\,\mathbf{B}^\top \tag{11}$$

be the singular-value decomposition of $\mathbf{V}$. Then, instead of the pseudo-inverse

$$\mathbf{V}(\mathbf{x})^+ = \mathbf{B}\,\Sigma^{-1}\,\mathbf{A}^\top \tag{12}$$

in (6), we let determinantal SARSA use the following "noisy" pseudo-inverse:

$$\mathbf{V}(\mathbf{x})^+ \approx \mathbf{B}\left((1-\varepsilon_t)\,\Sigma^{-1} + \varepsilon_t\,\mathbf{M}\right)\mathbf{A}^\top, \tag{13}$$

where $\mathbf{M}_{ij} = \delta_{ij}$. We gradually reduce the magnitude of the noise ε_t over the iteration t.

4 Experiments

In our experiments, we evaluate the performance of determinantal SARSA with and without the new method of avoiding poor local optimal. We conduct our experiments on the blocker task (Sallans and Hinton 2001, 2004; Heess et al. 2017; Sallans 2002), for which we have shown in Osogami and Raymond (2019) that Determinantal SARSA outperforms baseline methods (Sallans and Hinton 2001, 2004; Heess et al. 2017; Sallans 2002). We closely follow the instances considered in Heess et al. (2017) and Osogami and Raymond (2019). All of the experiments are carried out with Python implementation on a workstation having 48 GB of memory and a 4.0 GHz CPU.

In the blocker task, we control an agent team, consisting of three agents, in a collaborative manner, where the goal is to let one of the agents reach the end zone, while two blockers hinder the agents. The field is a grid of four rows and seven columns. The three agents start at uniformly random positions in the top row. The two blockers, each occupies three consecutive squares, start at uniformly random positions in the bottom row. At each time step, each agent can move one step in one of the four directions or stay. After all of the agents take actions, each blocker moves one step to the right or to the left if doing so can block an agent; otherwise, the blocker stays. If one of the agents reaches the end zone, the agent team receives $+1$ reward for that time step. Otherwise, the agent team incurs -1 reward per time step. See Heess et al. (2017) and Osogami and Raymond (2019) for more details about the exact settings. Note that the blocker task is performed on a fully observable environment. In Osogami and Raymond (2019), we also evaluate determinantal SARSA with stochastic policy tasks, where the environment is partially observable.

As we have done in Osogami and Raymond (2019), we represent the team action a_t by a $4 \times 7 = 28$ dimensional binary vector $\mathbf{x}$, where each element indicates whether an agent occupies a particular square after taking the team action a_t ($(\mathbf{x}_t)_i = 1$) or not. Likewise, we set $\mathbf{D}_t \equiv \mathbf{I}$ and $\alpha = 0$. Hyperparameters of determinantal SARSA are set as in Osogami and Raymond (2019).

Figure 2 shows the performance of determinantal SARSA without the use of the technique of avoiding poor local optima. Specifically, the average reward per step is evaluated for every 40,000 steps (and for every 4000 steps during the initial 40,000

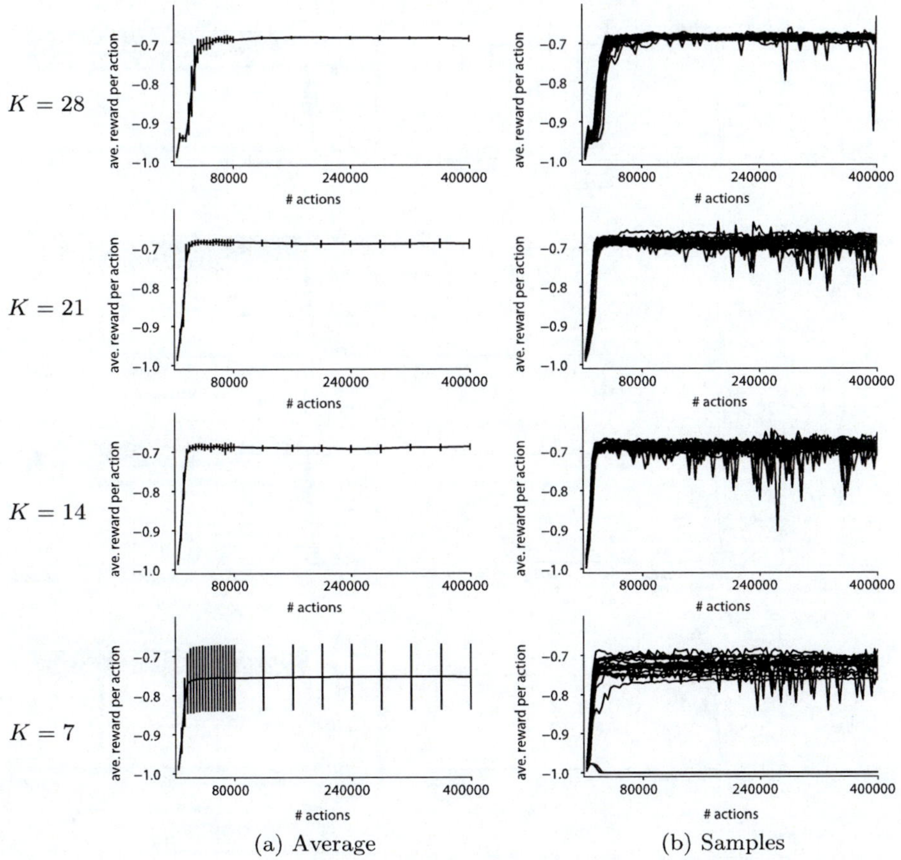

Fig. 2 Performance of determinantal SARSA with various ranks of kernels, where the rank K is indicated in the leftmost column. The panels in **(a)** show the mean and the standard deviation, over 20 runs, of the average reward per action. The panels in **(b)** show the average reward per action for each of the 20 runs

steps). The rank K of the kernel is varied as indicated in each row. Figure 2a shows the mean and the standard deviation, over 20 runs, of the average reward per action. Figure 2b shows the average reward per action for each of the 20 runs.

We can observe in Fig. 2a that reducing rank K has only a small impact on the average performance for $K \geq 14$. However, a significant degradation in performance

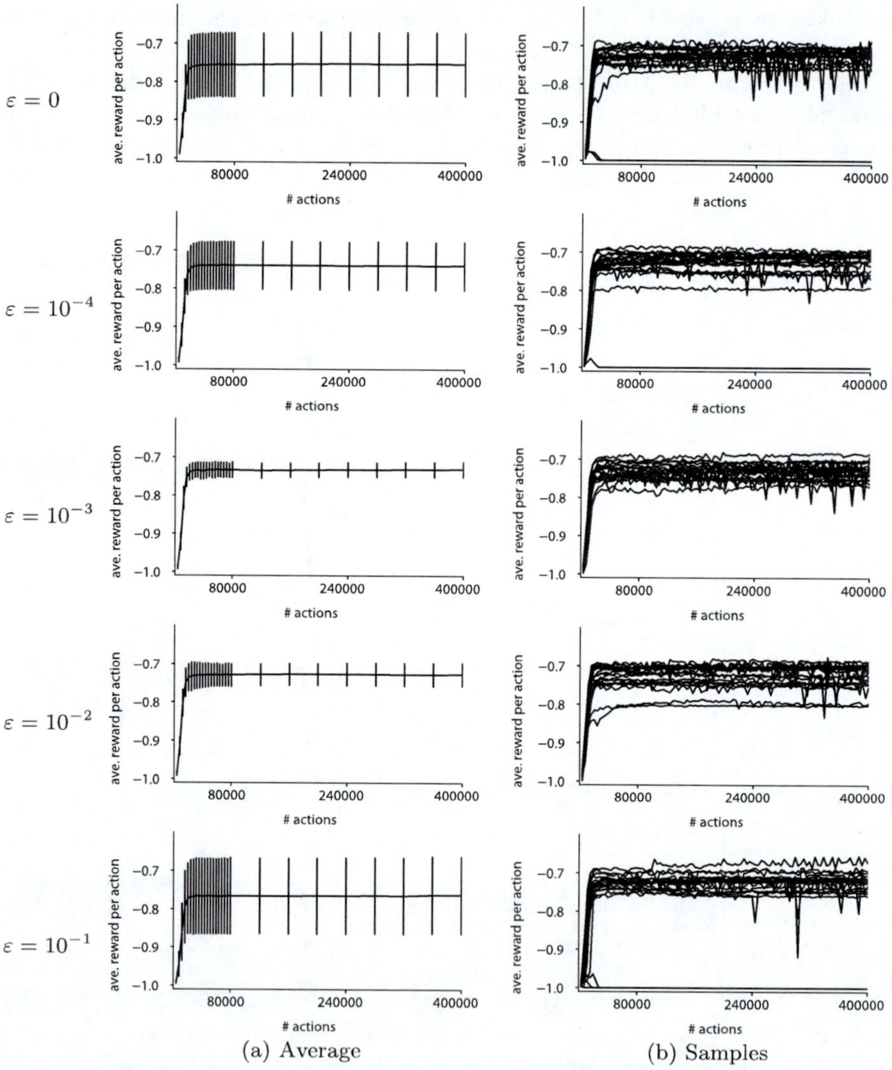

Fig. 3 Performance of determinantal SARSA with noisy initialization and pseudo-inverse when $K = 7$, where the magnitude ε of the noise at every 10,000-th iteration is indicated in the leftmost column. The panels in **(a)** show the mean and the standard deviation, over 20 runs, of the average reward per action. The panels in **(b)** show the average reward per action for each of the 20 runs

is observed with $K = 7$. Figure 2b shows that determinantal SARSA sometimes converges to poor local optima with $K = 7$. In particular, it has converged to the average reward of -1 (the lowest possible average reward) in two out of 20 runs.

Figure 3 shows the performance of determinantal SARSA with $K = 7$ when the technique of mitigating poor local optima is applied. Here, the magnitude of the noise at the t-th iteration (ε_t in (13)) is defined to be

$$\varepsilon_t = \varepsilon^{\frac{t}{10^4}}. \tag{14}$$

Namely, $\varepsilon_0 = 1$ and $\varepsilon_{10^4} = \varepsilon$, where various values of ε are tested as indicated in the leftmost column in Fig. 3. It suggests that determinantal SARSA can avoid convergence to poor local optima by the use of noisy gradient with appropriate magnitude of noise (specifically, $10^{-3} \le \varepsilon \le 10^{-2}$).

5 Conclusion

In Osogami and Raymond (2019), we have introduced determinantal SARSA, which uses the determinant of a matrix so that both diversity and relevance of team actions can be taken into account in reinforcement learning. Determinantal SARSA has been shown to substantially outperform existing methods proposed for coping with high-dimensional action space in multi-agent reinforcement learning. Determinantal SARSA can effectively deal with exponentially large team action space. When there are 2^N possible team actions, determinantal SARSA has at most $O(N^3)$ computational complexity and can have smaller complexity by assuming a low rank structure.

However, we find that determinantal SARSA with low-rank kernels can result in poor local optima. In this paper, we have proposed techniques of noisy initialization and noisy pseudo-inverse to avoid the poor local optima in determinantal SARSA. The results of numerical experiments support the effectiveness of the proposed techniques.

Acknowledgements This work was supported by JST CREST Grant Number JPMJCR1304, Japan.

References

Belabbas MA, Wolfe PJ (2009) On landmark selection and sampling in high-dimensional data analysis. Philos Trans R Soc: Math Phys Eng Sci 367:4295–4312

Gillenwater J (2014)Approximate inference for determinantal point processes. Ph.D. thesis

Hausknecht M, Stone P (2015) Deep recurrent Q-learning for partially observable MDPs. In: Sequential decision making for intelligent agents: papers from the AAAI 2015 fall symposium, pp 29–37

Heess N, Silver D, Teh YW (2013) Actor-critic reinforcement learning with energy-based policies. In: Proceedings of the 10th European workshop on reinforcement learning, vol 24., Edinburgh, Scotland, pp 45–58
Kang B (2013) Fast determinantal point process sampling with application to clustering. Adv Neural Inf Process Syst 26:2319–2327
Kulesza A, Taskar B (2011) k-DPPs: fixed-size determinantal point processes. In: Proceedings of the 28th international conference on machine learning, pp 1193–1200
Kulesza A, Taskar B (2012) Determinantal point processes for machine learning. Now Publishers Inc., Hanover, MA, USA
Osogami T (2017) Boltzmann machines for time-series. Technical report RT0980, IBM Research—Tokyo
Osogami T, Otsuka M (2015) Seven neurons memorizing sequences of alphabetical images via spike-timing dependent plasticity. Sci Rep 5:14149
Osogami T, Raymond R (2019) Determinantal reinforcement learning. In: Proceedings of the 33nd AAAI conference on artificial intelligence, pp 4659–4666
Qiao M, Xu RYD, Bian W, Tao D (2016) Fast sampling for time-varying determinantal point process. ACM Trans Knowl Discovery Data 11(Article No. 8)
Sallans B (2002) Reinforcement learning for factored Markov decision processes. Ph.D. thesis
Sallans B, Hinton GE (2001) Using free energies to represent Q-values in a multiagent reinforcement learning task. Adv Neural Inf Process Syst 13:1075–1081
Sallans B, Hinton GE (2004) Reinforcement learning with factored states and actions. J Mach Learn Res 5:1063–1088
Stewart GW (1980) The efficient generation of random orthogonal matrices with an application to condition estimators. SIAM J Numer Anal 17(3):403–409
Wachinger C, Golland P (2015) Sampling from determinantal point processes for scalable manifold learning. In: Proceedings of the international conference on information processing in medical imaging, pp 687–698

Surface Denoising Based on Normal Filtering in a Robust Statistics Framework

Sunil Kumar Yadav, Martin Skrodzki, Eric Zimmermann, and Konrad Polthier

1 Introduction

Surface denoising—generally being part of the preprocessing stage in the geometry processing pipeline—is designed to remove high-frequency noise corrupting a geometry. The noise generally arises from scanning or other acquisition processes. In contrast to smoothing, we are interested in preserving attributes and features of the geometry like edges and corners. Here, the difficulty lies in distinguishing these from noise, depending on the intensity of noise and the level of the attributes' details.

Denoising can therefore be considered as being part of the area of smoothing. It is used in all applications asking for a cleaned, i.e., noise-free, surface with the additional property of keeping features. But more importantly, it is recognized as being a major tool in the preprocessing stage of geometry processing. The reason is that—besides computer-designed models—the acquisition of real world models via 3D scanning processes unfortunately adds noise and outliers to the data due to mechanical limitations and sub-optimal surrounding conditions. These artifacts influence meshes and point sets alike and have to be removed to obtain a clean model for further use in different industry applications, e.g., scientific analysis, automotive, medical diagnosis, rendering, and other geometry processing algorithms like surface reconstruction, feature detection, computer-aided design, or 3D printing, see (Yadav et al. 2018b) for applications in medical diagnoses and (Botsch et al. 2010) for a variety of application scenarios.

S. K. Yadav · M. Skrodzki · E. Zimmermann · K. Polthier
Department of Computer Science and Mathematics, Freie Universität Berlin, Berlin, Germany

S. K. Yadav (✉)
Technical Development, Nocturne GmbH, Berlin, Germany
e-mail: sunil.yadav@fu-berlin.de

M. Skrodzki
Computer Graphics and Visualization, TU Delft, Delft, The Netherlands

J. Cheng et al. (eds.), *Proceedings of the Forum "Math-for-Industry" 2018*,
Mathematics for Industry 35, https://doi.org/10.1007/978-981-16-5576-0_6

A typical challenge arising in the denoising process is the decoupling of noise and features of a geometry. This is, because both are high-frequency components of the geometry in terms of the spectral setting. Other problems arise as noisy geometries include outliers, which are far away from the underlying ground truth. Furthermore, the amplitude of noise can be significant when compared to the feature size. To solve these problems, in both cases—for meshes and point sets—a variety of surface denoising algorithms have been published. These state-of-the-art methods can be categorized into:

1. One-stage methods, where noise components are removed by adjusting the vertex positions based on curvature information;
2. Two-stage methods, wherein the first stage, surface normals are filtered and then in the second stage, vertex positions are adjusted according to the filtered normals.

Two-stage methods are more effective in terms of feature preservation as well as noise removal and obtain minimum volume shrinkage compared to one-stage methods, see (Centin and Signoroni 2018; Yadav et al. 2018c, 2019). In the two-stage methods, surface normal filtering is the key part as it is responsible for both noise removal and feature preservation. Therefore, several procedures have been published for normal filtering. Each of these algorithms is effective in different aspects (like robustness against noise, feature preservation, or detection of outliers). However, there is no unified theoretical framework available in which we can discuss the benefits and drawbacks of the normal filtering algorithms and in which we can derive relations between these methods.

In this paper, we focus on this issue and introduce such a unified framework making use of robust statistics to derive relations between (both linear and nonlinear) state-of-the-art surface normal filtering methods. On the basis of these relations, we discuss the robustness of each algorithm against noise and its respective feature preservation capability. The presented framework can be used to provide pros and cons of published methods for the development of new algorithms. Furthermore, it can serve as a comparison possibility for such new procedures to state-of-the-art methods on a theoretically sound basis.

1.1 Notation

Throughout the whole paper, we will use the following notation. Let I, J, K denote index sets as subsets of $\mathbb{N}$. We consider a mesh $\mathcal{M} = (P, E, F)$ consisting of a set of points or vertices $P = \{p_i\}_{i \in I} \subset \mathbb{R}^3$ (which will be used in the point set setting as well), (undirected) edges E, and faces F. In general, we will assume that the mesh $\mathcal{M}$ or the point set P is corrupted by noise. The set of normals is given as $N = \{n_j\}_{j \in J} \subset \mathbb{S}^2$, with $\mathbb{S}^2$ the two-dimensional unit sphere in $\mathbb{R}^3$ and neighborhoods are labeled Ω_k for $k \in K$. Sometimes we only refer to the neighborhood by Ω and to its representatives by $p, q \in \Omega$ without further labels, to simplify the notation where it is unambiguous. The used type of neighborhood will get specified when necessary and receive a dedicated index set, as it further depends on the context, i.e., to which object (points, faces, ...) we are going to relate it. Consequently, normals and neighborhoods apply for faces and points depending on whether we discuss the mesh or point set setting. Let $|X|$ denotes the size of a set X and let $\|v\|$ as well as v^T be the Euclidean norm and the transpose of a vector $v \in \mathbb{R}^3$, respectively. A surface area or a vertex, both of high curvature (in comparison with the other elements of the geometry) will be referred to as a *feature* of the mesh or the point set, respectively.

1.2 Related Work

In the last two decades, many surface smoothing algorithms have been developed. Due to the large number of available methods, for a comprehensive overview we refer to (Botsch et al. 2010; Centin and Signoroni 2018). Here, we give a short overview of methods highly related to the robust statistics setting and of the most important state-of-the-art methods. As stated above, the removal of noise components is equivalent to the removal of high-frequency components. Here, the Fourier transform is a common tool, allowing efficient implementations of low-pass filters to cut off high frequencies. It has been generalized to manifold harmonics to be applicable to 2-manifold surfaces via the eigenfunctions of the Laplace–Beltrami operator of these surfaces. Its matrix representation encodes the *natural vibrations* of a triangle mesh in its eigenvectors and the *natural frequencies* in its eigenvalues, see (Taubin 1999, 2001a). One drawback is its cost for many applications as the eigenvector decomposition of the Laplace matrix is numerically challenging to compute; see (Vallet and Levy 2008).

A similar removal of high-frequency components can be achieved by utilizing the diffusion flow, which dampens high frequencies (instead of cutting them off) by a multiplication with a Gaussian kernel. It can be computed directly on the mesh, making it cheaper and hence more practical than the Fourier transform. Let $f(\mathbf{p}, t) : \mathbb{R}^{3|P|+1} \to \mathbb{R}$ be a given signal with $\mathbf{p} = (p_1, \ldots, p_{|P|})^T$. The diffusion equation:

$$\frac{\partial f(\mathbf{p}, t)}{\partial t} = \lambda \Delta f(\mathbf{p}, t) \tag{1}$$

describes the change of f over time by a scalar diffusion coefficient $\lambda \in \mathbb{R}$ multiplied with its spatial Laplacian Δf, which can be replaced by the Laplace–Beltrami operator on manifolds. As the discretization asks for small time steps to be numerically robust in the integration, the authors of (Desbrun et al. 2001) proposed an implicit time integration providing unconditional robustness even for large time steps. A smoothing procedure can be derived from this as update of the vertex positions p_i by a point-wise update scheme

$$\begin{aligned} p_i &\leftarrow p_i + h\lambda \Delta p_i, \\ \text{with } \Delta p_i &= -2\,H n_i, \end{aligned} \tag{2}$$

because the Laplace–Beltrami operator on vertices corresponds to the mean curvature. Hence, all vertices p_i move in the corresponding normal direction n_i by a magnitude regulated by the mean curvature H. This is known as the *mean curvature flow*, see (Desbrun et al. 2001).

The isotropic Laplacian has been extended by a data-dependent diffusion tensor yielding the anisotropic flow equation:

$$\frac{\partial f}{\partial t} = \text{div}[g_\sigma(\|\nabla f\|)\nabla f], \tag{3}$$

where f is a signal as in Eq. (1) and $g_\sigma(\cdot)$ is an edge stopping function (anisotropic weighting function), which is responsible for feature preservation with a user input parameter σ during denoising operations, see (Perona and Malik 1990; Clarenz et al. 2000). Further examples for the usage of the anisotropic diffusion equation can be found in (Bajaj and Xu 2003; Hildebrandt and Polthier 2004). The same concept is extended to the context of point set smoothing by Lange and Polthier (Lange and Polthier 2005) and to face normal filtering by Tasdizen et al. (Tasdizen et al. 2002).

Another set of denoising techniques consists of two-stage mesh denoising algorithms. Here, at the first stage, face normals are filtered and in the second stage vertex positions are updated according to the newly computed face normals, see (Taubin 2001b). Face normal filtering is performed by using several linear and nonlinear filters in order to preserve sharp features (Centin and Signoroni 2018; Yadav et al. 2018c; Yagou et al. 2002, 2003; Ohtake et al. 2002; Belyaev and Ohtake 2001) and vertex updates are performed by using the edge-face orthogonality (Sun et al. 2007).

Finally, there are several denoising methods utilizing bilateral filtering. It arose from image processing (Tomasi and Manduchi 1998) and uses a combination of two different weighting functions: a spatial kernel and a range kernel to preserve features and remove noise components. It got adapted to surface denoising for instance in (Fleishman et al. 2003), where the information of spatial distances and the local variation of vertex normal vectors are combined for denoising. Bilateral filters are extended for face normal filtering, where a range kernel (Gaussian function) is defined based on the normal differences in the neighborhood (Yadav et al. 2019; Zheng et al. 2011). A variation of bilateral filtering is also used extensively in mesh denoising in order to remove noise and retain sharp features (Jones et al. 2003; Zhang et al. 2015).

1.3 Face Normal Filtering Versus Vertex Position Filtering

Broadly, surface smoothing algorithms can be divided into two categories, direct vertex position filtering, which is also known as one-stage smoothing and two-stage filtering, which includes (face) normal filtering and vertex position updates as described above.

Most of the one-stage denoising algorithms (vertex position filtering) follow the concept of mean curvature flow, which is related to the Laplace–Beltrami operator and the mean curvature on the surface as shown in Eq. (2) and as discussed above. Basically, noise components are removed by minimizing the mean curvature on the surface, where the mean curvature is computed using the area gradient on the surface. Therefore, minimizing the curvature will result in minimizing the area, which will lead to volume shrinkage. This applies to most of the anisotropic and isotropic diffusion-based surface smoothing algorithms. These methods use vertex position filtering in their minimization. To illustrate this problem, Fig. 1a shows a noisy model and Fig. 1b shows the result obtained by using the mean curvature flow-based method of (Hildebrandt and Polthier 2004). More precisely, Fig. 1b shows two different surfaces, the original surface (green) and the denoised one (yellow). The difference between these two surfaces is visible due to volume shrinkage during the minimization.

On the other hand, in two-stage surface denoising, noise removal is performed based on the face normals. Basically, face normals are treated as signals on the vertices of the dual graph of the mesh with values in the unit sphere. The face normal denoising is generally performed by rotating the face normals on the unit sphere according to the weighted average of the corresponding neighbor face normals (see Eq. (5) for a formalization). In other words, for noise removal, we operate in the dual space of the mesh and minimize the variation of face normals. This operation does not involve the curvature minimization on the vertex positions. Therefore, in two-stage surface denoising algorithms, volume shrinkage is minimal, as shown in Fig. 1c, d.

Furthermore, in two-stage surface denoising, noise removal can be performed also on vertex normals (Fleishman et al. 2003) instead of face normals. However, in terms of sharp feature preservation, vertex normal filtering will not be as effective as face normal filtering because of the following reasons:

1. The vertex normals of a mesh are usually derived from face normals. Therefore, processing face normals will avoid the ill-posedness and increase the robustness of the algorithm.
2. At a sharp feature, the angle between vertex normals is smaller than the angle between the face normals. Therefore, face normals are more robust in feature preservation compared to vertex normals.

As shown in Fig. 1c, d, face normal filtering better preserves sharp features compared to vertex normal filtering methods. However, in the context of point set surfaces, face normals are not available and denoising has to be performed using vertex normals.

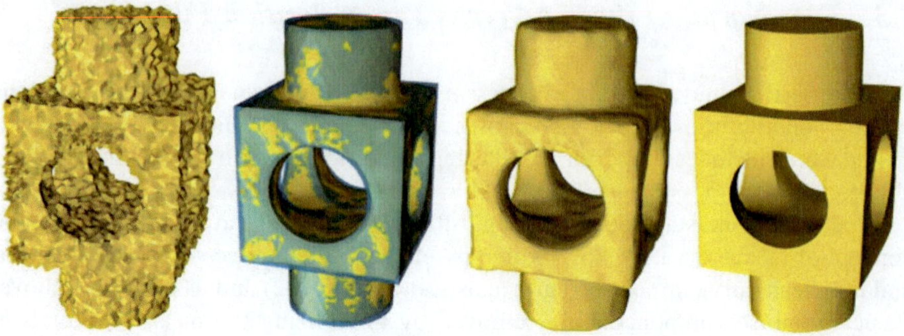

Fig. 1 A visual comparison between vertex position, vertex normal, and face normal filtering methods. **a** shows the noisy block model and **b** shows the denoised result of the method presented in Hildebrandt and Polthier (2004), based on mean curvature flow. More precisely, it shows two different surfaces, the original surface (green) and the denoised one (yellow). The difference between these two surfaces is visible due to volume shrinkage during the minimization. In contrast, **c, d** show the result of the face normal filtering methods (Fleishman et al. 2003) and (Yadav et al. 2018c), respectively, which do not suffer from volume shrinkage

1.4 Scope

From our discussion in the last section, it is clear that the two-stage surface denoising algorithms are robust and efficient in terms of noise removal and feature preservation. Therefore, in this paper, we will cover surface normal filtering (face normal in the context of mesh surfaces and vertex normals in the context of point set surfaces) in a robust statistics framework.

In the context of surface denoising, the most challenging task is to decouple sharp features from noise to treat them appropriately. Robust statistics is an efficient tool to identify the deviating substructures (*outliers*) from the bulk data. Here, we will treat features on the geometry as *outliers* because we want to deal with features differently compared to the non-feature areas. Based on this assumption, we derive relationships between different state-of-the-art methods for surface normal filtering using the concept of the robust error norm and its corresponding influence functions; see Sect. 2. We also discuss the robustness of these algorithms within the presented framework, see Sects. 3 and 4.

2 Robust Statistical Estimation

This paper is concerned with robust statistics handling outliers during statistical data modeling. The field of robust statistics has developed methods to handle outliers in the data modeling process, see (Mrázek et al. 2006). These methods describe the structure of best fitting the bulk of the data and identifying deviating substructures (outliers), see (Black and Rangarajan 1996). In this section, we translate the robust statistics framework to the setting of surface denoising. As explained above, surface denoising is a preprocessing operation in many geometry processing algorithms, which removes noise components and retains sharp features. In the robust statistics framework, surface features can be seen as outliers and methods from robust statics can identify these, which in turn can be treated differently for feature-preserving surface denoising, see (Yadav et al. 2019). As stated in the notation, we consider both a face and a vertex of the surface mesh to be a *feature*, respectively, if the corresponding normals of its neighbors have a high variation. Note that this is also the case for noisy faces and vertices, but not for outliers as they will not have a close neighborhood.

As reasoned in Sect. 1.4, we focus on two-stage mesh denoising algorithms. Recall that—as it is mentioned in Sect. 1.1—the surface $\mathcal{M}$ is corrupted by noise. Therefore, the vertices P and face normals N contain noise components, too. Let us first assume that the noise-free surface is represented by $\hat{\mathcal{M}}$ with $\hat{P}$ and $\hat{N}$ its vertices and face normals, respectively. The noisy and noise-free face normals can be related by:

$$n = \hat{n} + \eta, \tag{4}$$

where η is a random variable representing the noise corrupting the surface. If η is a zero-mean Gaussian random variable and the surface is flat, then the denoised face normals can be computed by minimizing the following L_2 error to compute the mean:

$$E(\hat{n}) = \sum_{n\in\Omega} \|\hat{n} - n\|^2, \qquad \hat{n} = \frac{1}{|\Omega|}\sum_{n\in\Omega} n. \tag{5}$$

However, in real-life scenarios, the noise η is not always normally distributed and surfaces have sharp features, which can be seen as outliers. Therefore, in the following we will aim at computing an approximation $\tilde{n}$ of $\hat{n}$. To deal with this complicated situation, we use robust error norms, which lead to the theory of M-estimators, see Sect. 2.1 for details. An M-estimator of a face normal from noisy normals can be obtained as the minimum of the following error functional:

$$E_\sigma(\tilde{n}) = \sum_{n\in\Omega} \rho_\sigma\left(\|\tilde{n} - n\|\right), \tag{6}$$

where $\rho_\sigma(\cdot) : \mathbb{R} \to \mathbb{R}$ is a loss function and commonly called ρ-function or error norm (Black and Rangarajan 1996; Black et al. 1998; Durand and Dorsey 2002) and the quantity σ is a user input. See Table 1 for different choices for ρ_σ. To minimize the effect of outliers, the loss function should not grow rapidly. To see the growing speed of the robust error norm $\rho_\sigma(\cdot)$, its derivative is computed, which is referred to as influence function ($\psi_\sigma(\cdot)$) in robust statics (Winkler et al. 1998). Thus, the loss function and influence function are related as follows

$$\rho'_\sigma(x) =: \psi_\sigma(x), \tag{7}$$

where for convenience, let us put $x := \|\tilde{n} - n\|$.

During mesh denoising, at sharp features, the effect of the influence function should be minimal. The input parameter x will be related to features, i.e., to the variation of normals. Therefore, when $x \to \infty$, the influence function should be zero, that is

$$\lim_{x\to\infty} \psi_\sigma(x) = 0.$$

In our setting, feature values (x) are basically defined by the variation of normals, which is measured by the differences between the neighboring normals n_j and the central normal n_i. However, these differences cannot approach infinity practically as $n_i, n_j \in \mathbb{S}^2$ for all $i, j \in I$. Therefore, the above equation indicates that for bigger values of x the influence function should be diminished.

Equation (6) can be extended to take into account spatial weights in local neighborhoods using the following formulation:

$$E_{\sigma,\sigma_d}(\tilde{n}) = \sum_{n\in\Omega} \rho_\sigma\left(\|\tilde{n} - n\|\right) f_{\sigma_d}(d), \tag{8}$$

where the function $f_{\sigma_d}(d)\colon \mathbb{R} \to \mathbb{R}$ is an isotropic weighting factor, which takes the spatial distance d between the considered geometry elements as the input argument and is responsible for smoothing out high-frequency components of the geometry. The term σ_d controls the width of the spatial kernel and generally depends on the resolution (sampling density) of the given geometry. In case of mesh denoising, the distance is computed between the centroid of neighboring faces and the processed central face. For point set denoising, the term d is computed between neighboring vertices and the processed central vertex.

Throughout the whole paper, concerning the error functionals, we are going to ignore constant factors in the arguments for both the isotropic (σ_d) and the anisotropic (σ) case. This is to focus on the qualitative differences between the presented methods rather than on smaller variations.

2.1 M-estimators

M-estimators are collections of different robust error norms to handle outliers. Any estimator defined by Eq. (6) is called an "M-estimator." The name comes from the generalized maximum likelihood concept, which can be deduced from Eq. (6), when $-\rho_\sigma(x)$ is the likelihood function. Then, minimizing the energy $E_\sigma(\cdot)$ of Eq. (6) will be equivalent to the maximum likelihood estimate (Chu et al. 1998; Hampel et al. 2005). As motivated above, in general, the robust estimators should have the following two properties:

1. The error norm $\rho_\sigma(x)$ should not grow rapidly.
2. The influence function $\psi_\sigma(x) = \rho'_\sigma(x)$ should be bounded.

For an efficient mesh denoising procedure, the influence function should be a *re-descending function*, i.e., $\psi_\sigma(x) \to 0$ when $x \to \infty$. In this case, the corresponding error norm $\rho_\sigma(x)$ is called *re-descending influence error norm* (Hampel et al. 2005).

In general, surface normal (i.e., face and vertex normal) filtering is performed by computing weighted averages of neighboring normals; see Eq. (11). The weighting functions are vital for feature-preserving normal filtering, and they can be either linear or nonlinear. Here, we will formulate the relationship between weighting function, robust error norm, and the corresponding influence function.

From Eq. (3), we know that the anisotropic diffusion is controlled by an edge stopping function, which is represented by $g_\sigma(x)$. In this paper, we termed it as anisotropic weighting function. Equation (6) can be minimized using gradient descent to update the surface normal:

$$n^{t+1} = n^t + \lambda \nabla E_\sigma(x) = n^t + \lambda \sum_{n \in \Omega} \nabla \rho_\sigma(\|\tilde{n} - n\|), \tag{9}$$

where t is the iteration number and λ represents the step size. Here, ρ_σ is interpreted as a concatenation, taking the norm of a vector as argument, while the norm receives $(\tilde{n}) \in \mathbb{R}^3$ as argument. The complete function then maps from $\mathbb{R}^3$ to $\mathbb{R}$. The differentiation let us consider the gradient of ρ_σ as a natural generalization of the derivative in the one-dimensional case. Following the reasoning of Jones et al. (2003), also adapted by Zheng et al. (2011), we adapt the procedure introduced in Tomasi and Manduchi (1998) for signal processing to the context of mesh processing by feeding the normal distance x—as defined above—into the error norm ρ_σ and a spatial distance into the spatial weighting function f_σ. This analogy motivates us to analyze the following well-established relation from signal processing [consider for a specific derivation (Black and Rangarajan 1996, Sects. 4.1 and 5.3) and more generally (Hampel et al. 2005; Huber 1981),

$$g_\sigma(x) = \frac{\rho'_\sigma(x)}{x} =: \frac{\psi_\sigma(x)}{x}. \tag{10}$$

Applications of this relation in image and geometry processing can be found in Jones et al. (2003), Black et al. (1998), Durand and Dorsey (2002).

The weighting function $g_\sigma(x)$ should capture the anisotropic behavior of the mesh or the point set, respectively, and should be chosen based on the above relations in the robust statistics framework. Table 1 consists of several well-known M-estimators with their robust error norms, their influence functions, and their corresponding anisotropic weighting functions.

Equation (5) shows an example of an estimator with a quadratic error norm ($\rho_\sigma(x) = x^2$). This norm grows rapidly, and its influence function ($\psi_\sigma(x) = 2x$) is unbounded (non-re-descending) as shown in Table 1. Therefore, the quadratic estimator is very sensitive to outliers and not useful in feature-preserving mesh denoising.

Table 1 M-estimators

Error norm $\rho_\sigma(x)$	Error norm $\rho_\sigma(x)$	Influence function $\psi_\sigma(x) = \rho'_\sigma(x)$	Weighting function $g_\sigma(x) = \frac{\psi_\sigma(x)}{x}$
◇ L_2-norm (Black et al. 1998), independent of σ, $\rho_\sigma(x) = x^2$		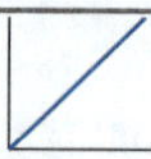	
◇ Truncated L_2-norm (Black and Rangarajan 1996) $\rho_\sigma(x) = \begin{cases} x^2 & \lvert x\rvert < \sqrt{\sigma} \\ \sigma & otrw. \end{cases}$	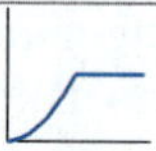		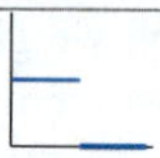
◇ L_1-norm (Hampel et al. 2005), independent of σ, $\rho_\sigma(x) = \lvert x\rvert$	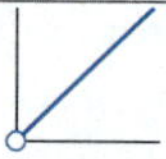	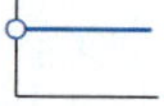	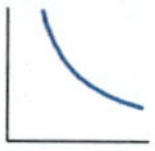
◇ Truncated L_1-norm (Hampel et al. 2005) $\rho_\sigma(x) = \begin{cases} \lvert x\rvert & \lvert x\rvert < \sigma \\ \sigma & otrw. \end{cases}$		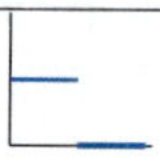	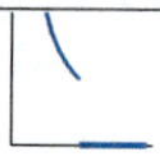
◇ Huber's minimax (Huber 1981) $\rho_\sigma(x) = \begin{cases} \frac{x^2}{2\sigma} + \frac{\sigma}{2} & \lvert x\rvert < \sigma \\ \lvert x\rvert & otrw. \end{cases}$	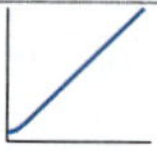		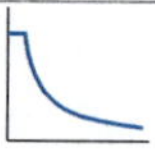
◇ Lorentzian-norm (Black et al. 1998) $\rho_\sigma(x) = \log\left[1 + \frac{1}{2}\left(\frac{x}{\sigma}\right)^2\right]$		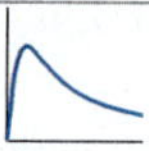	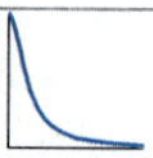
◇ Gaussian norm (Black and Rangarajan 1996) $\rho_\sigma(x) = 1 - e^{\left(-\frac{x^2}{\sigma^2}\right)}$	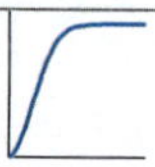	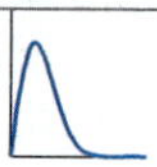	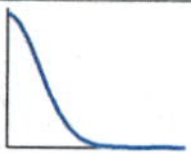
◇ Tukey's norm (Beaton and Tukey 1974) $\rho_\sigma(x) = \begin{cases} \frac{x^2}{\sigma^2} - \frac{x^4}{\sigma^4} + \frac{x^6}{3\sigma^6} & \lvert x\rvert < \sigma \\ \frac{1}{3} & otrw. \end{cases}$		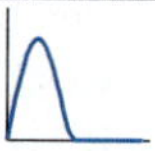	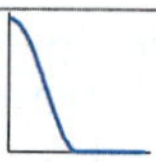

The quadratic error norm can be truncated in order to convert it into a re-descending influence error norm. The second row of Table 1 shows the truncated quadratic error norm that has a re-descending influence function $\psi_\sigma(x)$ with a bounded error norm $\rho_\sigma(x)$. However, the behavior of $\psi_\sigma(x)$ is linearly increasing within the range of the user input σ, which is not desired for feature preservation.

As shown in Table 1, the L_1 error norm ($\rho_\sigma(x) = |x|$, third row) and Huber's minimax error norm (fifth row) do not have re-descending influence functions even though they are bounded by a nonzero constant value. These two perform better in terms of separating outliers compared to the (truncated) quadratic error norm.

The other error norms listed in Table 1, which include the truncated L_1 error norm as well as the Lorentzian, Gaussian, and Tukey's norms have re-descending influence functions. Among all re-descending influence error norms, the truncated L_1

and Tukey's error norm cut off the influence function's response strictly while the other norms have a nonzero influence function on a larger interval.

3 Face Normal Filtering in the Robust Statistics Framework

In this section, we will discuss state-of-the-art methods for face normal filtering utilizing the robust statistics framework and M-estimators as described above. Based on the relationship between the robust error norm, the influence function, and the weighting function as established in Eq. (10), we will discuss the robustness and effectiveness of state-of-the-art methods for removing noise and preserving features.

The face normals N of a triangulated mesh $\mathcal{M}$ can be seen as graph signals on the graph induced by the dual mesh of $\mathcal{M}$ with values in the unit sphere. The centroid of each face f_i is denoted by c_i, which can be treated as the vertex position on the dual mesh. In general, the filtered face normal $\tilde{n}_i$ corresponding to a noisy face normal n_i can be computed using the following equation:

$$\tilde{n}_i = \frac{1}{\omega} \sum_{j \in \Omega_i} g_\sigma \left(\left\| n_i - n_j \right\|^2 \right) f_{\sigma_d}(\left\| c_i - c_j \right\|^2) n_j, \tag{11}$$

where $\omega = \left\| \sum_{j \in \Omega_i} g_\sigma(\left\| n_i - n_j \right\|^2) f_{\sigma_d}(\left\| c_i - c_j \right\|^2) n_j \right\|$ ensures $\tilde{n}_i$ to be of unit length. The term Ω_i represents the mesh neighborhood around the ith triangle, which can be combinatorial or a geometrical disk of some (user-defined) radius. The above equation represents a general formula for face normal filtering and follows the error functional presented in Eq. (8). The efficiency of this approach heavily depends on the choice of the weighting functions $g_\sigma(\cdot)$ and $f_{\sigma_d}(\cdot)$.

In the following, we will present several state-of-the-art approaches for these choices. The listed algorithms use different input arguments for the robust error functionals. Common choices are the Euclidean distance of normals $\left\| n_i - n_j \right\|$, the angle between two normals $\angle(n_i, n_j)$, or the quantity $\arccos(n_i \cdot n_j)$. We will stick to the notation used in the respective original paper in the following discussion. However, note that these input arguments are related. In particular, we obtain

$$\cos(\angle(n_i, n_j)) = \frac{n_i \cdot n_j}{\|n_i\| \, \|n_j\|} = n_i \cdot n_j \Rightarrow \angle(n_i, n_j) = \arccos(n_i \cdot n_j).$$

by the Euclidean scalar product because all normals considered are of unit length. Furthermore, (by the law of cosines) it is

$$\begin{aligned}\left\|n_i - n_j\right\|^2 &= \|n_i\|^2 + \left\|n_j\right\|^2 - 2\cdot\|n_i\|\cdot\left\|n_j\right\|\cdot\cos(\angle(n_i, n_j))\\ &= 2 - 2\cos(\angle(n_i, n_j))\\ \Rightarrow \angle(n_i, n_j) &= \arccos\left(1 - \frac{\left\|n_i - n_j\right\|^2}{2}\right).\end{aligned}$$

3.1 Unilateral Normal Filtering

Unilateral normal filtering performs noise removal from noisy normals using a single anisotropic kernel function. From our setup in Eq. (8), it is clear that the unilateral normal filtering algorithms are using $g_\sigma(x)$ as anisotropic weighting function while the spatial filter will be equal to one, i.e., $f_{\sigma_d}(d) \equiv 1$. These methods are effective against low intensity of noise and enhance sharp features. However, they are not robust against moderate or high levels of noise because of the unavailability of the spatial filter $f_{\sigma_d}(d)$.

(a) Belyaev and Ohtake (2001) introduce nonlinear diffusion of face normals to enhance the features of the geometry. Their algorithm uses the following weighting function:

$$g_\sigma(x) = \exp\left(-\frac{x^2}{\sigma^2}\right). \tag{12}$$

This weight is a nonlinear function, and the input argument is encoding the directional curvature. It is given as

$$x = \frac{\angle(n_i, n_j)}{d},$$

where $\angle(n_i, n_j)$ denotes the angle between n_i and n_j, the term $d = \left\|c_i - c_j\right\|$ represents the distance between the centroids (as presented above) of the central face and its neighboring face, and $n_i, n_j \in N$ are faced normals of the central face and its neighboring face, respectively. The term σ is a user input to better adapt the algorithm to the given geometry. It is chosen based on the amount of noise, curvature, and the resolution of the geometry. The directional curvature x measures the similarity between neighboring normals. In the robust statistics framework, by using Eq. (10), we can deduce the used error norm as

$$\rho_\sigma(x) = \int_0^x x' g_\sigma(x')dx' = \frac{\sigma^2}{2}\left(1 - \exp\left(-\frac{x^2}{\sigma^2}\right)\right). \tag{13}$$

Similarly, the influence function can be derived as

$$\psi_\sigma(x) = x g_\sigma(x) = x \exp\left(-\frac{x^2}{\sigma^2}\right), \qquad \lim_{x\to\infty} \psi_\sigma(x) = 0. \tag{14}$$

The above two equations indicate that this algorithm applies the Gaussian error norm (second last row of Table 1), which has a re-descending influence function and makes the algorithm robust against outliers. However, the spatial smoothing function $f_{\sigma_d}(\cdot)$ is not used in this algorithm, which reduces the robustness of the algorithm against significant noise.

(b) Yagou et al. (2002) apply mean and median filtering to face normals. Mean filtering of normals is performed by simply uniformly averaging neighboring normals. Therefore, the anisotropic weighting function $g_\sigma(x) \equiv 1$ leads to an error norm and influence function of

$$\rho_\sigma(x) = \int_0^x x' g_\sigma(x') dx' = x^2 \qquad \text{and} \qquad \psi_\sigma(x) = x g_\sigma(x) = x, \tag{15}$$

respectively. From the equation above, it is clear that mean filtering follows the quadratic error norm ($\rho_\sigma(x) = x^2$, $g_\sigma(x) = 1$) (the first row in Table 1) and it has an unbounded influence function ($\lim_{x\to\infty} \psi_\sigma(x) = \infty$), which makes the algorithm sensitive to outliers and produces feature blurring. This method uses the triangle area as a weighting function, i.e., in the notation of Eq. (8), it computes $f_{\sigma_d}(d)$ for a given face f_i as area(f_i). However, this makes the algorithm only insensitive to irregular sampling.

On the other hand, median filtering is estimated using the L_1 error norm (Hampel et al. 2005). Therefore, the corresponding error norm and influence function can be derived as

$$\rho_\sigma(x) = |x| \qquad \text{and} \qquad \psi_\sigma(x) = \rho'_\sigma(x) = \begin{cases} 1 & |x| \neq 0 \\ \text{undefined} & x = 0. \end{cases} \tag{16}$$

By using the relation from Eq. (10), the anisotropic weighting function can be written as

$$g_\sigma(x) = \frac{\psi_\sigma(x)}{x} = \begin{cases} \frac{1}{|x|} & |x| \neq 0 \\ \text{undefined} & x = 0. \end{cases} \tag{17}$$

In this algorithm, the input x is given by the Euclidean distance of the neighboring normal $n_j \in N$ to the central normal n_i, i.e., $x = \|n_i - n_j\|$. The L_1-norm is better compared to the quadratic error norm in terms of robustness to outliers. However, the corresponding influence function is not re-descending (see Table 1) and produces a constant value for outliers.

Weighted median filtering is applying a spatial weighting function to provide higher weights to closer points compared to distant points; see (Yagou et al. 2002). This weighting function is truncating the effect of local neighboring faces. Therefore, the weighted median follows a truncated L_1-norm and its corresponding influence function can be derived as

$$\psi_\sigma(x) = \rho'_\sigma(x) = \begin{cases} 0 & |x| < \sigma \\ \text{sign}(x) & 0 < |x| \leq \sigma \\ \text{undefined} & x = 0 \end{cases}. \tag{18}$$

By using the relation from Eq. (10), the anisotropic weighting function can be written as

$$g_\sigma(x) = \frac{\psi_\sigma(x)}{x} = \begin{cases} 0 & |x| < \sigma \\ \frac{\text{sign}(x)}{x} & 0 < |x| \leq \sigma \\ \text{undefined} & x = 0 \end{cases}. \tag{19}$$

The truncated L_1-norm has a re-descending influence function, which enhances the feature preservation capability of the algorithm compared to mean and median filtering.

From the influence functions of the L_1-norm and the truncated L_1-norm, it is clear that these norms are capable of feature preservation during the process of face normal filtering. However, these influence functions and their corresponding anisotropic weighting functions are not well-defined at $x = 0$, which is not desirable.

(c) Huber (1981) proposes a slight modification of the weighting function before mentioned to overcome the issue of not being well-defined at $x = 0$. He suggests

$$\rho_\sigma(x) = \begin{cases} \frac{x^2}{2\sigma} + \frac{\sigma}{2} & |x| < \sigma \\ |x| & \text{otrw.} \end{cases}. \tag{20}$$

This modified error norm is commonly known as Huber's minimax norm (see fifth row in Table 1). The corresponding influence and anisotropic weighting functions can be derived as

$$\psi_\sigma(x) = \begin{cases} \frac{x}{\sigma} & |x| < \sigma \\ \text{sign}(x) & \text{otrw.} \end{cases}, \qquad g_\sigma(x) = \begin{cases} \frac{1}{\sigma} & |x| < \sigma \\ \frac{\text{sign}(x)}{x} & \text{otrw.} \end{cases}. \tag{21}$$

The above equation indicates that Huber's minimax norm has a re-descending influence function and has a well-defined anisotropic weighting function. This norm is widely used in image processing applications but has—to the best of our knowledge—not been used for face normal filtering yet and is therefore not included in Table 2.

(d) Yadav et al. (2018c) introduced a face normal filtering technique using a box filter as the anisotropic weighting function

$$g_\sigma(x) = \begin{cases} 1 & |x| < \sigma \\ 0.1 & \text{otrw.} \end{cases}, \qquad \text{with} \qquad x = \angle(n_i, n_j), \tag{22}$$

where $\angle(n_i, n_j)$ denotes the angle between the central normal n_i and it neighboring normal n_j. The corresponding error norm and influence function can be derived as

$$\rho_\sigma(x) = \int_0^x x' g_\sigma(x') dx' = \begin{cases} x^2 & |x| < \sigma \\ 0.1(x^2 + 9\sigma^2) & \text{otrw.} \end{cases}, \tag{23}$$

$$\psi_\sigma(x) = x g_\sigma(x) = \begin{cases} x & |x| < \sigma \\ 0.1x & \text{otrw.} \end{cases}. \tag{24}$$

From the above error norm and influence function, we can see that this filtering is using an error norm quite similar to the truncated quadratic error norm (see second row in Table 1) for the computation of the element-based normal voting tensor. The corresponding influence function is neither bounded nor re-descending, but the outlier effect will be quite minimal. This is because of the downscaling of the argument in the influence function for bigger x. Therefore, the algorithm is able to preserve sharp features. However, it is less robust against high noise intensities because of the non-re-descending and unbounded influence function.

(e) Shen et al. (2004) introduced the fuzzy vector median-based surface smoothing algorithm, which is quite similar to the algorithm of (Belyaev and Ohtake 2001) (explained in paragraph **a)** in the beginning of this section). The anisotropic weighting function $g_\sigma(x)$ is a Gaussian function as given in Eq. (12) and the input x is given as

$$x = \left\| n_j - n_{vd} \right\|,$$

where n_j represents neighboring normals to the processed central face f_i and the term n_{vd} performs *vector directional median filtering* on the normal vectors including the central normal n_i. Vector directional median filtering is an extension of *median filtering* for multivariate data, see (Trahanias and Venetsanopoulos 1993), and can be computed as

$$n_{vd} = \underset{n}{\text{argmin}} \sum_{j \in \Omega_{vd}} \angle(n, n_j), \tag{25}$$

where $\angle(n, n_j)$ denotes the angle between n and n_j and the set $\Omega_{vd} = \Omega_i \cup \{i\}$ consists of indices of the neighbor normals n_j together with the index i of the central normal n_i.

The corresponding influence function will be re-descending as shown in Eqs. (13) and (14). The input argument of $g_\sigma(x)$ is the Euclidean difference between the neighboring normals and their median. This method performs well in terms of feature preservation but is not robust during noise removal because of the unavailability of the spatial filter. As it is clear from Eqs. (3), the anisotropic weighting function $g_\sigma(x)$ is similar to the edge stopping function in the diffusion process.

(f) Tasdizen et al. (2002) apply—based on the relationship between bilateral filtering and nonlinear diffusion (Barash 2002)—the diffusion of face normals for filtering by using the Gaussian function as anisotropic weighting function. Curvature information is used as input x in this algorithm. Similar to the method of (Belyaev and Ohtake 2001), from Eqs. (12), (13), and (14), it can be derived that this method also follows the Gaussian error norm and has a bounded, re-descending influence function, which helps preserving sharp features. However, due to the unavailability of the spatial filter, this algorithm is not robust against significant noise.

(g) Centin et al. (2018) also introduce a face normal diffusion method using the following anisotropic weighting function

$$g_\sigma(x) = \begin{cases} 1 & |x| < \sigma \\ \frac{\sigma^2}{(\sigma - x)^2 + \sigma^2} & \text{otrw.} \end{cases}, \qquad \text{where} \qquad x = \kappa \cdot \ell_{avg}. \tag{26}$$

The term κ represents curvature information computed at each face by averaging the curvature at the corresponding vertices and ℓ_{avg} represents the average edge length computed over the entire geometry. The corresponding influence function can be derived as

$$\psi_\sigma(x) = x g_\sigma(x) = \begin{cases} x & |x| < \sigma \\ \frac{x\sigma^2}{(\sigma - x)^2 + \sigma^2} & \text{otrw.} \end{cases}. \tag{27}$$

The above influence function is bounded and re-descending, which makes this algorithm effective in terms of feature preservation. This method falls somewhere between the Lorentzian error norm (decaying of $g_\sigma(x)$ for $x \geq \sigma$) and Huber's minimax error norm (constant $g_\sigma(x)$ for $x < \sigma$). Due to the absence of a spatial filter, this algorithm is not robust against high intensities of noise.

3.2 *Bilateral Normal Filtering*

Bilateral normal filtering is one of the most effective and robust approaches for denoising of normals. In contrast to unilateral normal filtering, the weighting function in bilateral normal filtering consists of two different Gaussian kernels. As above, one kernel carries the anisotropic nature and is commonly known as range filter (we termed it anisotropic weighting function $g_\sigma(x)$) while the other kernel is known as spatial kernel (given as $f_{\sigma_d}(d)$ in Eq. (8)) and is isotropic in nature.

(a) Zheng et al. (2011) define these kernels as:

$$g_\sigma(x) = \exp\left(-\frac{x^2}{2\sigma^2}\right) \qquad \text{and} \qquad f_{\sigma_d}(d) = \exp\left(-\frac{d^2}{2\sigma_d^2}\right), \tag{28}$$

where σ_d is the average distance between neighboring faces and the central face. The input arguments x and d are defined as:

$$x = \left\| n_i - n_j \right\| \qquad \text{and} \qquad d = \left\| c_i - c_j \right\|,$$

where c_i and c_j are the centroids of the central face f_i and the neighboring face f_j, respectively.

In the robust statistics framework, our main focus is the anisotropic weighting function $g_\sigma(x)$, its corresponding error norm, and the corresponding influence function because $g_\sigma(x)$ is responsible for feature preservation. From Eqs. (12), (13), and (14), it is clear that the method of (Zheng et al. 2011) has a re-descending influence function (second last row of Table 1). Thereby, this algorithm is capable of preserving sharp features effectively and removes noise better compared to the algorithms mentioned above because of the utilized spatial filter $f_{\sigma_d}(d)$.

(b) Zhang et al. (2015) describe a procedure of guided mesh normal filtering following the Gaussian error norm and uses the same spatial filter as the method of (Zheng et al. 2011) presented above. The guided mesh normal is based on a joint bilateral filter, where an anisotropic weighting function (range kernel) works on the guidance signal. That is, the input variable x is defined as:

$$x = \left\| G_i - G_j \right\|, \tag{29}$$

where G_i and G_j are the guidance normals, which are computed by averaging similar normals in the respective neighborhood.

(c) Yadav et al. (2019) introduce a bilateral normal filtering using the following anisotropic weighting function:

$$g_\sigma(x) = \begin{cases} \frac{1}{2}\left[1 - \left(\frac{x}{\sigma}\right)^2\right]^2 & |x| \le \sigma \\ 0 & \text{otrw.} \end{cases}, \qquad \text{where} \qquad x = \left\| n_i - n_j \right\|. \tag{30}$$

The above function is known as Tukey's biweight function (Beaton and Tukey 1974). The spatial filter $f_{\sigma_d}(d)$ is a Gaussian function similar to that used in the method of (Zheng et al. 2011) as described above. In the robust statistics framework, the corresponding influence function and error norm can be derived as

$$\psi_\sigma(x) = x g_\sigma(x) = \begin{cases} \frac{x}{2}\left[1 - \left(\frac{x}{\sigma}\right)^2\right]^2 & |x| < \sigma \\ 0 & \text{otrw.} \end{cases}, \tag{31}$$

$$\rho_\sigma(x) = \int_0^x x' g_\sigma(x') dx' = \begin{cases} \frac{x^2}{\sigma^2} - \frac{x^4}{\sigma^4} + \frac{x^6}{3\sigma^6} & |x| < \sigma \\ \frac{1}{3} & \text{otrw.} \end{cases}. \tag{32}$$

From the influence function and error norm, it is clear that Tukey's biweight function is more robust compared to the Gaussian function in terms of feature preservation because it strictly cuts off outliers with respect to the user-chosen parameter σ. Also, the Gaussian spatial filter helps to remove noise components effectively.

4 Point Set Surface Denoising in the Robust Statistics Framework

In this section, we will shift our focus slightly. Instead of an input mesh $\mathcal{M}$, we will now consider a point set sample (PSS) of a surface as input. Thus, we are only given vertices $P = \{p_i\}_{i \in I} \subseteq \mathbb{R}^3$ with corresponding normals $N = \{n_i\}_{i \in I}$, i.e., compared to the above we cannot use edges to induce connectivity between the vertices nor can we use the area of faces as weighting terms in the filtering process.

Despite these challenges, a multitude of procedures and algorithms has been proposed for the denoising of PSS. This is mostly due to two advantages of PSS over meshes. First, point sets are often the raw output of 3D acquisition devices and processes. Thus, if an algorithm is available to work on a PSS, it can be directly—possibly even on site—applied to the acquired data. Second, as there is no connectivity information in the point set, no such data has to be stored, which amounts to significantly lower storage costs compared to meshes. Furthermore, no topological problems—like non-manifold edges or fold-overs—and no numerical problems—like slivers—are introduced as the PSS only gives an implicit handle on the underlying surface geometry.

In the following, we will focus on adaptations of face normal filtering algorithms from meshes to point sets as well as on original methods proposed directly in the PSS setting. Note that any method on point sets can easily be applied to the meshed setting by simply disregarding the edge and face connectivity information.

4.1 Unilateral Normal Filtering

As for meshes, we will first focus on unilateral normal filtering procedures. These do not use a specific spatial filter, i.e., $f_{\sigma_d}(d) \equiv 1$. This makes them less robust against moderate or high levels of noise.

(a) Öztireli et al. (2009) introduced a modification of the moving least squares (MLS) procedure (Alexa et al. 2003) aiming at the integration of feature preservation into the MLS pipeline. Their core objective is an iterative minimization and can be understood as iterative trilateral filtering, as it makes use of three types of weights. The first one is isotropic in nature and appears as C^3 continuous polynomial approximation of the Gaussian, i.e.,

$$f_i(p) = \left(1 - \frac{\|p - p_i\|^2}{h_i^2}\right)^4 \tag{33}$$

where the argument p is some point (not necessarily from P), as the objective is an implicit, signed distance function. The value h_i is a weight adapting the local density, chosen within a range from 1.4 to 4 as experimentally evaluated by the authors (Öztireli et al. 2009). For the second weighting term—using the height over an estimated hyperplane at p and thus capturing both isotropic and anisotropic quantities—the authors discuss M-estimators and include the Gaussian error norm and its respective Gaussian error weight, see Eq. (12), into their optimization problem. The arguments are

$$d = y_i - \tilde{\eta}^{k-1}(p_i) \qquad \text{and} \qquad \sigma_d = \frac{h_i}{2},$$

with y_i the heights of the samples p_i taken over the local least-squared best fitting hyperplane, and $\tilde{\eta}^{k-1}$ the corresponding local approximation. The value for σ_d is set fix throughout the whole paper by the authors. The third and final weighting terms are anisotropic and make use of a Gaussian function with arguments

$$x = \left\| \nabla \eta^k(p) - n_i \right\| \qquad \text{and} \qquad \sigma \in \mathbb{R},$$

where η is an implicit, signed distance function as main objective, p some point at which we want to evaluate the function η, n_i the normal at sample point p_i, and σ a parameter that regulates the sharpness where typical choices range from 0.5 up to 1.5. This last weighting term penalizes the deviation of normals when we reach sharp features. The influence function and error norm are of Gaussian nature and are derived in Eqs. (14) and (13). The assembled combination yields a robust implicit surface definition via MLS, which can represent both smooth surface patches and sharp features and was coined robust implicit MLS (RIMLS). Similar to Method (Belyaev and Ohtake 2001), this algorithm is capable of retaining and enhancing sharp features. However, the unavailability of a spatial filter $f_{\sigma_d}(d)$ makes the algorithm less effective against moderate and high levels of noise.

(b) Mattei and Castrodad (2016) start their paper with the assertion that the principal component analysis (PCA) operation for the estimation of local reference planes is not robust. They proceed to construct a moving robust PCA (MRPCA). Their main ingredient of interest in the given context is a minimization problem, which makes use of anisotropic weights determined via the Gaussian weight function as given in Eq. (12) with arguments

$$x = \arccos(n_i \cdot n_j) \qquad \text{and} \qquad \sigma \in \mathbb{R},$$

where n_i, n_j are the unit normals at the considered point p_i and at one of its neighbors p_j (with a k-nearest neighborhood utilized). Furthermore, σ is a bandwidth parameter affecting the reconstruction of sharp features. The authors propose values of $\sigma \in (\pi/12, \pi/6)$. Using this anisotropic weight function yields the Gaussian error norm along with its re-descending influence function as given in Eqs. (14) and (13). Similar to (Belyaev and Ohtake 2001), this algorithm is capable of retaining and enhancing sharp features. However, the unavailability of a spatial filter $f_{\sigma_d}(d)$ makes the algorithm less effective against moderate and high levels of noise.

4.2 Bilateral Normal Filtering

We will now turn to bilateral normal filtering procedures for PSS. These use two different weighting kernels. As for meshes, one kernel carries the anisotropic nature while the other one of isotropic behavior.

(a) Li et al. (2009) presented one of the first approaches applying bilateral filtering to PSS. The authors first estimate the likelihood ℓ_i that a given sample point $p_i \in P$ is close to the underlying surface geometry. They propose to compute ℓ_i based on the MLS technique of (Alexa et al. 2003). The normal denoising utilizes the bilateral filtering scheme, which includes a Gaussian weighting (following Eq. (12)) as a spatial filter $f_{\sigma_d}(d)$ with the following input arguments in the isotropic setting

$$d = \|p_i - p_j\| \qquad \text{and} \qquad \sigma_d = \frac{r}{2},$$

and another Gaussian weighting function $g_\sigma(x)$ in the anisotropic setting with following input arguments

$$x = \arccos(n_i \cdot n_j) \qquad \text{and} \qquad \sigma \in \mathbb{R},$$

the latter chosen to be the standard deviation of the normal variation given in x. Here, r is the radius of the enclosing sphere of the geometric neighborhood Ω_i. Observe that the values presented here differ from those given in (Li 2009), because we adjust them to fit the Gaussian given in Eq. (12). Lastly, the closeness of the point p_i to the underlying surface, measured by ℓ_i, the feature intensity, and the bilateral filtering for normals are used in a final sample point filtering step to remove noise from the PSS. The mentioned method follows the Gaussian error norm similar to the bilateral normal filtering of (Zheng et al. 2011). As shown in Eq. (14), the applied anisotropic weighting function $g_\sigma(x)$ has a re-descending and bounded influence function, which makes the algorithm robust in terms of feature preservation and also the availability of the spatial filter $f_{\sigma_d}(d)$ ensures the effectiveness toward different levels of noise.

(b) Zheng et al. (2017) proposed a four-stage method for point set denoising. It consists of sharp feature detection, multiple normals computation, guided normal filtering, and point updating. Concerning the feature detection, the authors provide a two-step procedure: feature candidate detection and feature point selection. The former is to find the global feature structure and utilizes the framework of robust statistics. Namely, after a first computation of normals using PCA, the normal similarity is evaluated via the Gaussian weight function, see Eq. (12), with arguments

$$x = \|n_i - n_j\| \qquad \text{and} \qquad \sigma \in \mathbb{R},$$

with a user-given angle-threshold σ, which ranges from 0.05 to 0.3 in the experiments of the authors, n_i the normal at the considered point and n_j the normal at one of its neighbors, while using the k-nearest neighbors as neighborhood notion. In contrast to the single normal used in the normal similarity described above, the authors of (Zheng et al. 2017) attach bundles—a multitude of normals—to every point. A comparable approach is then chosen to estimate averaged normals utilizing spatial weights evaluated once more via the Gaussian weight function (12) with arguments

$$d = \|p_i - p_j\| \qquad \text{and} \qquad \sigma_d \in \mathbb{R},$$

with σ_d ranging from 0.1 to 0.5 in the authors' experiments. Finally, both weightings are combined in the actual bilateral normal filtering. This method is an extension of guided mesh normal filtering (Zhang et al. 2015), which we have mentioned in Eq. (29). From the explanation for guided mesh normal filtering in Sect. 3.2, it is clear that this method also follows the Gaussian error norm along with a bounded and re-descending influence function and has similar robustness in terms of feature preservation and noise removal. The computation of guided normals makes this algorithm slightly better compared to bilateral normal filtering.

(c) Park et al. (2013) proposed a three-staged point set filtering approach including feature detection, normal re-calculation, and a point position update. Their feature detection tensor, adaptive sub-neighborhood, and point update all use the Gaussian weighting function given in Eq. (12), where for the first two, the arguments are of anisotropic nature given as

$$x = \sqrt{s^2 + c\kappa^2} \qquad \text{and} \qquad \sigma \in \mathbb{R},$$

with a prescribed constant c, σ set by the authors to the neighborhood range, which is 4δ with δ the arithmetic mean of all distances of the points to their closest neighbors respectively. The value s represents the arc-length on the tangent plane and κ the curvature obtained by the circle, which goes through both the center point p_i and its considered neighbor p_j and which is also tangent to the attached normals n_i and n_j. These normals are calculated via an initial normal estimation following (Hoppe et al. 1992). To compute the feature detection tensor, the method uses a Gaussian function as the anisotropic weighting, which has a re-descending influence function ψ_σ and a derived Gaussian error norm ρ_σ as given in Eqs. (14) and (13), respectively. In terms of feature sensitivity, it will be as effective as MRPCA. However, this algorithm is not robust against moderate and high levels of noise.

(d) Digne and de Francis (2017) proposed an extension of the bilateral filtering on meshes to points via a parallel implementation of (Fleishman et al. 2003) using points. The whole procedure consists of a point update using non-oriented normals and utilizes Gaussian weights, Equation (12), twice, with isotropic

$$d = \|p_i - p_j\| \qquad \text{and} \qquad \sigma_d = \frac{1}{3}r,$$

and anisotropic arguments

$$x = \left|n_i \cdot (p_j - p_i)\right| \qquad \text{and} \qquad \sigma = \frac{1}{3}r',$$

with user-given radii r and r'. If these are not given, the authors use a heuristic and set $r = \ell\sqrt{20/|P|}$, where ℓ denotes the size of the bounding box and $|P|$ the number of vertices. The values σ_d and σ are set to be equal in this case. The point p_i is the one considered to be updated and p_j represents one of its neighbors within a geometrical neighborhood Ω_i. The weights determined by f_{σ_d} measure the spatial distance, and those by g_σ evaluate the distance of neighbors to the plane spanned by the point p_i and its normal. As the weights are of Gaussian nature, we can derive the influence function and Gaussian error norm given in Eqs. (14) and (13). In terms of feature preservation and noise removal, this algorithm will be as effective as bilateral normal filtering (Zheng et al. 2011) as both of them are using same robust error norm with a slightly different input argument.

(e) Zheng et al. (2018) propose an iterative two-staged denoising algorithm which—in contrast to most methods—smooths out smaller features while preserving larger ones. The iterative normal filtering (with initial normals obtained via (Hoppe et al. 1992) and the following point position update (solved iteratively via gradient descent) make use of the Gaussian weighting, Equation (12), with the isotropic arguments

$$d = \left\|p_i - p_j\right\| \qquad \text{and} \qquad \sigma_d \in \mathbb{R}$$

and the anisotropic arguments

$$x = \left\|n_i - n_j\right\| \qquad \text{and} \qquad \sigma \in \mathbb{R},$$

where $\sigma_d \in [0.01, 0.5]$ and $\sigma \in [0.1, 0.5]$ given in the authors' experiments, p_i the considered point, p_j representing its neighbor (k-nearest neighbors are used), and n_i, n_j the respective normals. Consequently, the evaluation is similar and on the one hand uses spatial distances of points while on the other hand using closeness of normals. The used Gaussian weights yield the influence function and Gaussian error norm given in Eqs. (14) and (13), which make this algorithm robust in terms of feature preservation and noise removal. One of the key benefits of this algorithm is that by adjusting the parameter σ, different levels of features can be smoothed out effectively. An even more robust version, utilizing the same weighting terms as given above, is discussed in (Yangxing et al. 2019).

(f) Yadav et al. (2018a) offer an extension of (Yadav et al. 2018c) to point sets. The proposed iterative scheme consists of the following three stages: normal filtering, feature detection, and vertex update. The first two make use of a similar box filter as given in Eq. (22), here given as

$$g_\sigma(x) = \begin{cases} 1 & x \leq \sigma \\ 0 & \text{otrw.} \end{cases}$$

with input arguments

$$x = \arccos(n_i \cdot n_j) \qquad \text{and} \qquad \sigma \in \mathbb{R},$$

where n_i, n_j are unit-length normals and σ is an angle threshold for the neighbor selection (chosen by the user). The deviation from the weighting defined in (Yadav et al. 2018c) is because vertex normals are more sensitive to noise compared to face normals. Similar to the influence function and error norm derived in Eqs. (24) and (23), the anisotropic weights given above yield an influence function of

$$\psi_\sigma(x) = x g_\sigma(x) = \begin{cases} x & |x| < \sigma \\ 0 & \text{otrw.} \end{cases}$$

and an error norm of

$$\rho_\sigma(x) = \int_0^x x' g_\sigma(x') dx' = \begin{cases} x^2 & |x| < \sigma \\ 0 & \text{otrw.} \end{cases} .$$

The latter is a version of the truncated quadratic error norm, see the second row of Table 1. In contrast to (Yadav et al. 2018c), the influence function is both bounded and re-descending ($\psi \to 0$ when $x \to \infty$). The impact of outliers is therefore kept small as it scales down for larger arguments x and feature preservation is yielded. However, the performance of this algorithm is not optimal in the presence of moderate and high levels of noise due to the unavailability of a spatial filter $f_{\sigma_d}(d)$.

Discussion: Local versus Global Weighting Note that out of the methods for point set surface denoising presented here, only (Öztireli et al. 2009) utilizes a local vertex-based weight σ_d. In contrast, methods (Li 2009; Zheng et al. 2017; Digne and de Franchis 2017; Zheng et al. 2018) use global weighting terms σ_d. While localized terms can capture features on a finer level, they are harder to calibrate than global parameters. Furthermore, an implicit assumption of many algorithms is a noisy but uniformly dense sampling as input. Handling non-uniform densities requires additional work, see (Skrodzki et al. 2018). Finally, if the features of the input geometry are of comparable size, a global parameter is sufficient to capture them while still removing noise. Hence, most algorithms reduce to simple global parameters.

5 Experiments and Results

In this section, we present experimental results regarding the state-of-the-art methods as listed in the previous sections, which are using different robust error norms. We have chosen two different models (CAD and CAGD) with different levels of noise. Figure 2 shows the Nicola model corrupted with a moderate level of Gaussian noise (standard deviation $\sigma_n = 0.2\ell_e$, where ℓ_e is the average edge length). Using this model, we show the capability of feature preservation with the usage of different error norms. As shown in Fig. 2, the L_2-norm is not effective in terms of feature preservation (blurred eye region) because of the linear influence function and also as it is not bounded. The truncated L_2-norm preserves features in the eye region better compared to the L_2-norm as it has a truncated linear influence function. Figure 2e, f shows the outputs of using the Gaussian norm without and with spatial filter, respectively. The Gaussian error norm has a re-descending influence function, which makes the algorithm more effective compared to the L_2 related norms. The spatial filter is helping to remove noise effectively (eye and nose regions). Huber's minimax (Fig. 2e) and the Gaussian error norm (Fig. 2g) have quite similar outputs as they have re-descending influence functions and do not use spatial filters. Figure 2h shows the output of using Tukey's error norm, which has a sharper cut-off in the influence function compared to the Gaussian error norm. Therefore, feature preservation is better compared to other norms mentioned and the spatial filter is helping to remove noise components effectively.

Figure 3 shows the robustness of the mentioned norm against high level of noise. The Fandisk model is corrupted with a Gaussian noise ($\sigma_n = 0.3\ell_e$) in random direction. As it is shown, L_2 and Huber's minimax norms are able to remove the noise components effectively but feature preservation is not effective. In case of the Gaussian error norm, the spatial filter removes different components of noise including low-frequency ripples. However, the truncated L_2-norm is able to remove low-frequency components by introducing an additional processing step (binary optimization) in the pipeline. The algorithm (Yadav et al. 2019) uses Tukey's error norm, which helps to preserve features effectively and the spatial filter removes the noise components.

6 Conclusion

In this paper, we unified state-of-the-art methods for normal filtering in surface denoising using the robust statistics framework. We discussed different M-estimators, which are the main tools of robust statistics. These tools are defined by a robust error norm and a corresponding influence function, respectively. Based on the properties of the influence function (bounded and re-descending) and of the anisotropic weighting function, we discussed the robustness of state-of-the-art methods in terms of feature preservation and feature enhancement (see Table 2). Furthermore, we have shown that the introduction of spatial filters along with anisotropic filters will improve the robustness of the algorithm in terms of noise removal. The robust statistics framework not only provides a platform to bring new insight into the field of surface denoising and clarify the relation between different methods in the field. It can also be used for new methods to combine the advantages of the known filtering techniques. The application of robust statistics is not limited to surface denoising, and it can be used effectively in other areas of the field of geometry processing. Corresponding applications of this powerful tool are left as further research.

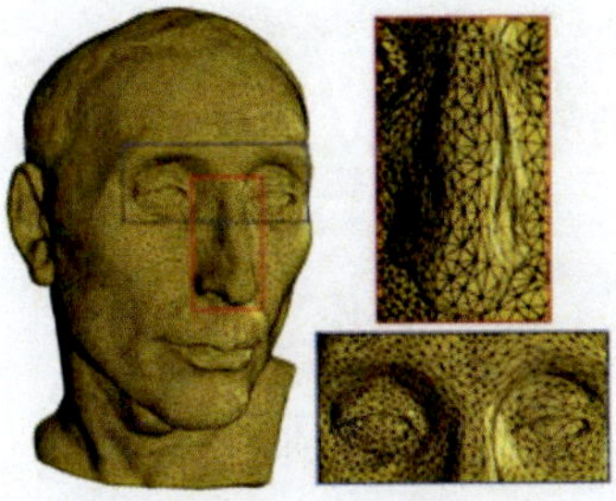

(a) Original

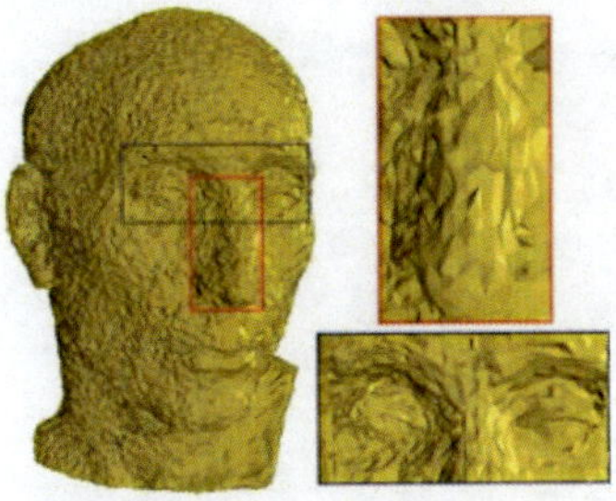

(b) Noisy

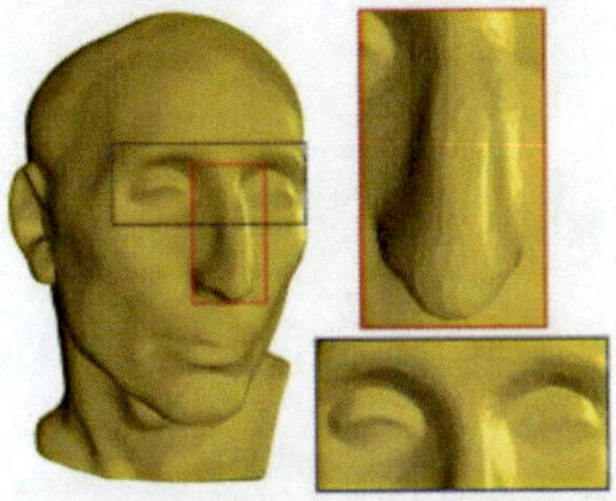

(c) L_2-norm (Yagou et al. 2002)

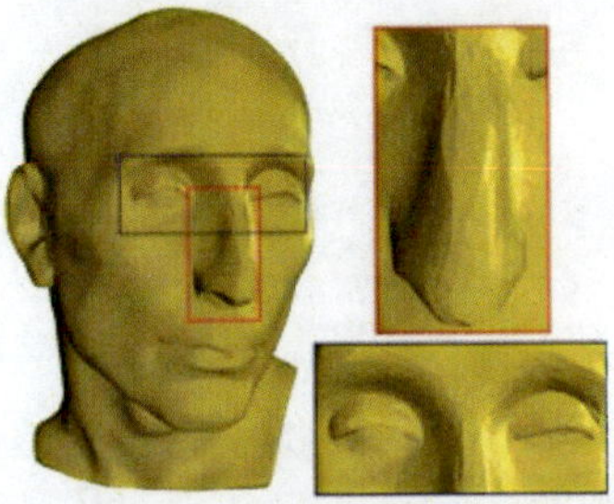

(d) Truncated L_2-norm (Yadav et al. 2018c)

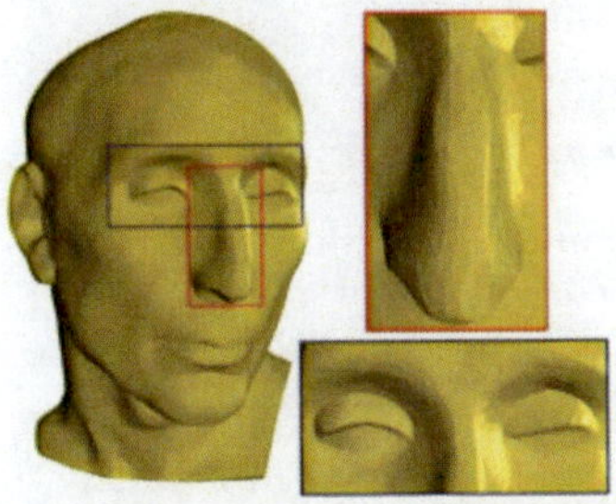

(e) Gaussian-norm (Belyaev and Ohtake 2001)

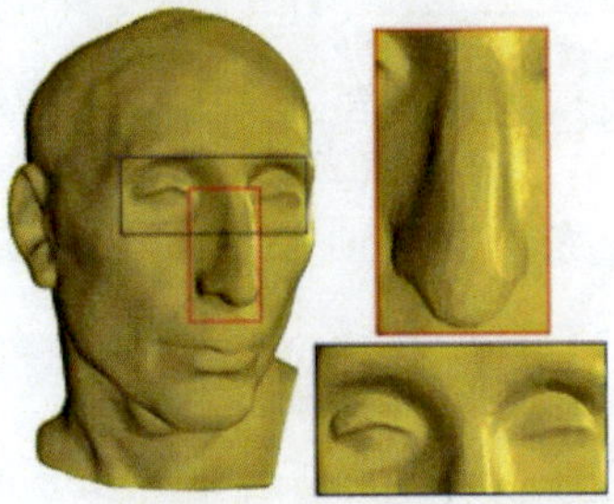

(f) Gaussian-norm with spatial filter (Zheng et al. 2011)

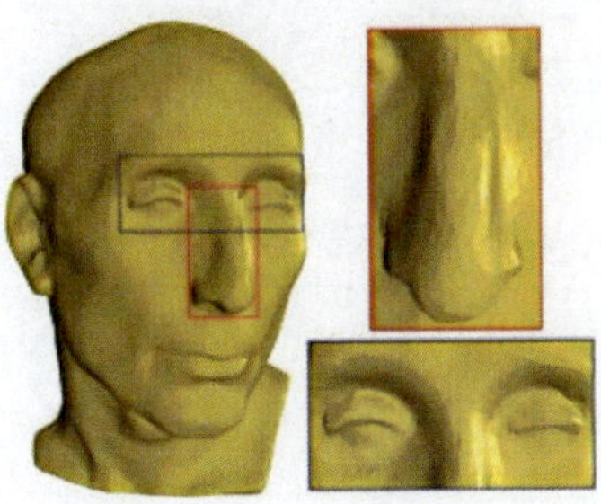

(g) Huber's minimax (Centin and Signoroni 2018)

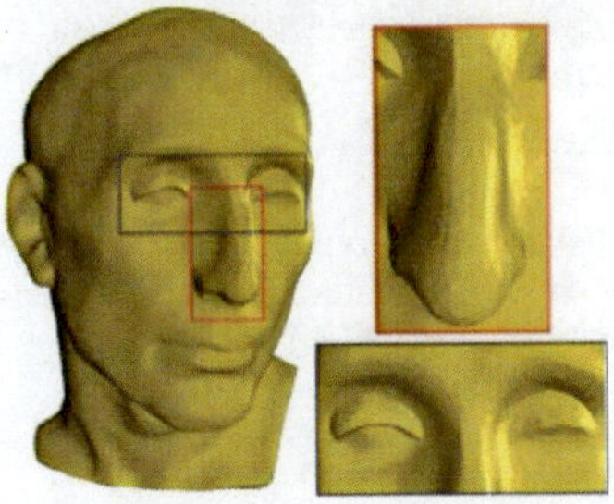

(h) Tukey's-norm (Yadav et al. 2019)

Fig. 2 Nicola model corrupted with a Gaussian noise ($\sigma_n = 0.2l_e$) in random direction. Images **c**–*h* show the results produced by state-of-the-art methods, which are using different robust error norms (see Table 1)

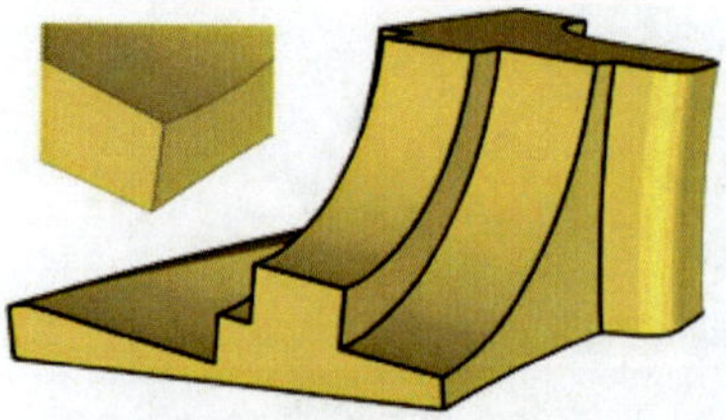

(a) Original

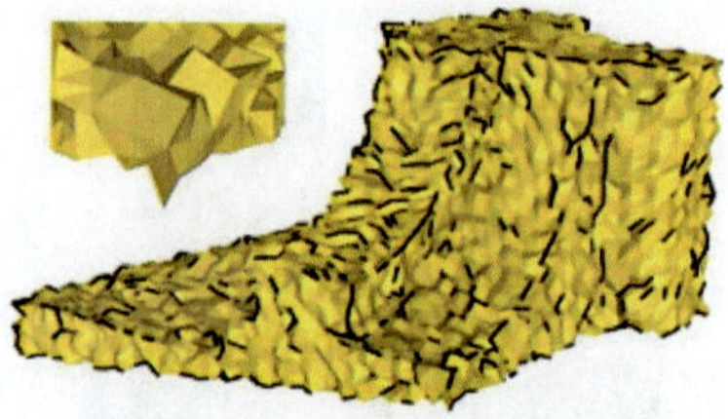

(b) Noisy

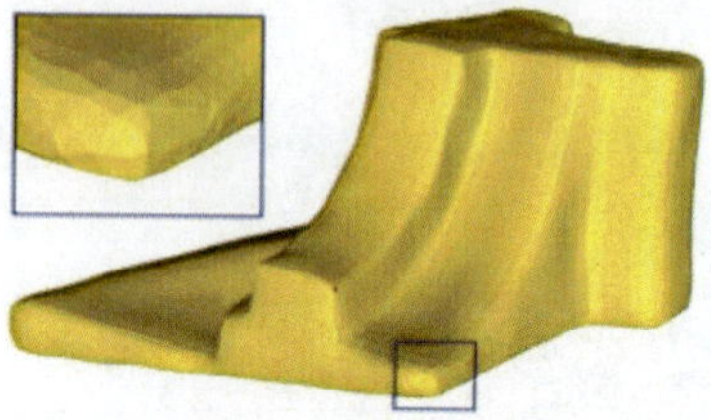

(c) L_2-norm (Yagou et al. 2002)

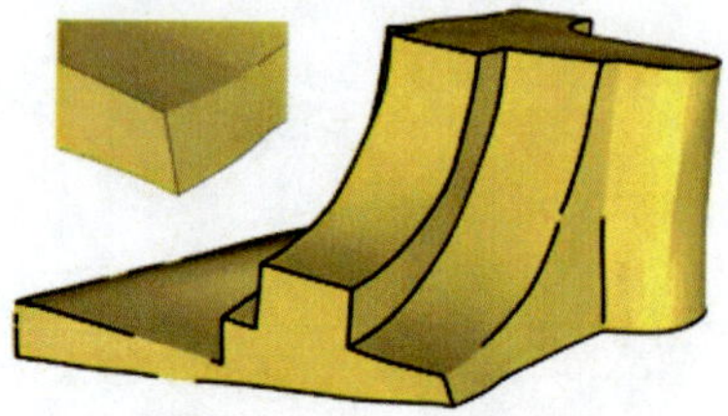

(d) Truncated L_2-norm (Yadav et al. 2018c)

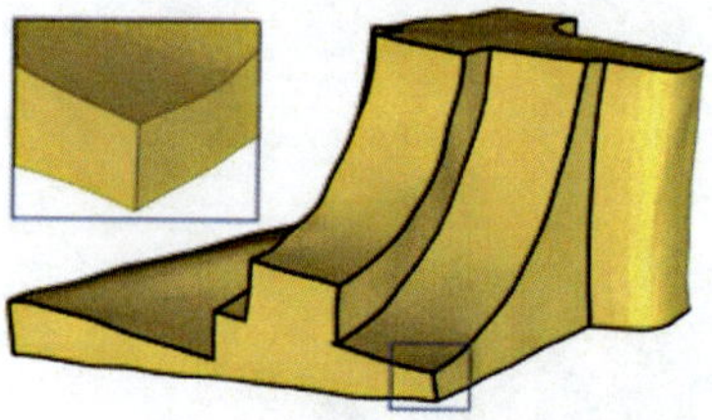

(e) Gaussian-norm (Belyaev and Ohtake 2001)

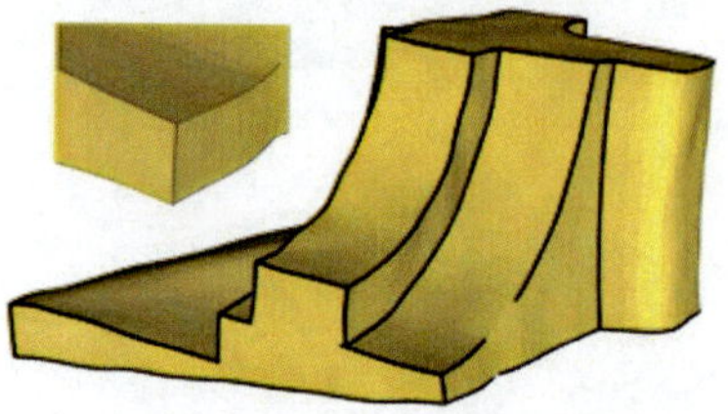

(f) Gaussian-norm with spatial filter (Zheng et al. 2011)

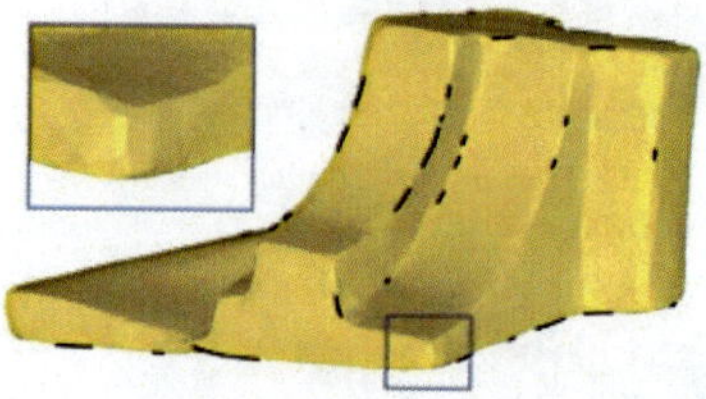

(g) Huber's minimax (Centin and Signoroni 2018)

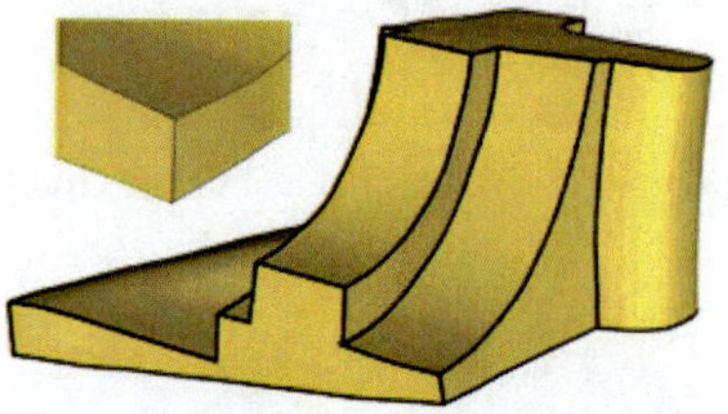

(h) Tukey's-norm (Yadav et al. 2019)

Fig. 3 Fandisk model corrupted with a Gaussian noise ($\sigma_n = 0.3\ell_e$) in random direction. **c**–*h* show the results produced by state-of-the-art methods, which are using different robust error norms (see Table 1). The black curve highlights sharp edge information in the geometries and is detected using a dihedral angle threshold of $\theta = 70°$

Table 2 Overview on the discussed methods. For each method, we present the authors, year, citation, which input is processed (PSS or meshes), what error norm is used and whether a spatial weighting is applied. Furthermore, we collect the assessments from the above sections how the different methods perform in terms of feature preservation and noise removal

Method	Section	Input	Error norm	Spatial weights	Feature preservation	Noise removal
Belyaev and Ohtake (2001)	3.1 a	Mesh	Gaussian	No	Good	Ok
Yogou et al. (2002)	3.1 b	Mesh	L_1 and L_2	No	Ok	Ok
Yadav et al. (2018c)	3.1 d	Mesh	Truncated L_2	No	Good	Ok
Shen and Barner (2004)	3.1 e	Mesh	Gaussian	No	Good	Ok
Tasdizen et al. (2002)	3.1 f	Mesh	Gaussian	No	Good	Ok
Centin and Signoroni (2018)	3.1 g	Mesh	Huber's minimax⋆	No	Excellent	Ok
Zheng et al. (2011)	3.2 a	Mesh	Gaussian	Gaussian	Good	Good
Zhang et al. (2015)	3.2 b	Mesh	Gaussian	Gaussian	Good	Good
Yadav et al. (2019)	3.2 c	Mesh	Tukey's	Gaussian	Excellent	Good
Öztireli (2009)	4.1 a	PSS	Gaussian	Gaussian	Good	Good
Mattei and Castrodad (2016)	4.1 b	PSS	Gaussian	No	Good	Ok
Li et al. (2009)	4.2 a	PSS	Gaussian	Gaussian	Good	Good
Zheng et al. (2017)	4.2 b	PSS	Gaussian	Gaussian	Good	Good
Park et al. (2013)	4.2 c	PSS	Gaussian	No	Good	Ok
Digne and Franchis (2017)	4.2 d	PSS	Gaussian	Gaussian	Good	Good
Zheng et al. (2018)	4.2 e	PSS	Gaussian	Gaussian	Good	Good
Yadav et al. (2018a)	4.2 f	PSS	Truncated L_2	No	Good	Ok

⋆ The error norm used in method (Centin and Signoroni 2018) is not equivalent to Huber's minimax. However, the utilized weighting term closely resembles the function $g_\sigma(x)$ of Huber's minimax, see Table 1 and the discussion in Sect. 3.1g

Acknowledgements This research was supported by the DFG Collaborative Research Center TRR 109, "Discretization in Geometry and Dynamics," the Berlin Mathematical School, the Einstein Center for Mathematics Berlin, Nocturne GmbH, and the German National Academic Foundation. The authors would like to thank the anonymous reviewer for many helpful suggestions and comments on how to improve the paper.

References

Alexa M, Behr J, Cohen-Or D, Fleishman S, Levin D, Silva CT (2003) Computing and rendering point set surfaces. IEEE Trans Visual Comput Graph 9(1):3–15

Bajaj CL, Xu G (January, 2003) Anisotropic diffusion of surfaces and functions on surfaces. ACM Trans Graph 22(1):4–32, Association for Computing Machinery, New York, NY, USA. ISSN: 0730-0301. https://doi.org/10.1145/588272.588276

Barash D (2002) Fundamental relationship between bilateral filtering, adaptive smoothing, and the nonlinear diffusion equation. IEEE Trans Pattern Anal Mach Intell 24(6):844–847

Beaton AE, Tukey JW (1974) The fitting of power series, meaning polynomials, illustrated on band-spectroscopic data. Technometrics 16(2):147–185

Belyaev AG, Ohtake Y (2001) Nonlinear diffusion of normals for crease enhancement. In: Vision geometry X, vol 4476, pp 42–48. International Society for Optics and Photonics

Black MJ, Rangarajan A (1996) On the unification of line processes, outlier rejection, and robust statistics with applications in early vision. International Journal of Computer Vision 19(1):57–91

Black MJ, Sapiro G, Marimont DH, Heeger D (1998) Robust anisotropic diffusion. IEEE Trans Image Process 7(3):421–432

Botsch M, Kobbelt L, Pauly M, Alliez P, Lévy B (2010) Polygon mesh processing. Peters

Centin M, Signoroni A (2018) Mesh denoising with (geo)metric fidelity. IEEE Trans Visual Comput Graph 24(8):2380–2396

Chu CK, Glad IK, Godtliebsen F, Marron JS (1998) Edge-preserving smoothers for image processing. J Am Stat Assoc 93(442):526–541

Clarenz U, Diewald U, Rumpf M (2000) Anisotropic geometric diffusion in surface processing. In: Proceedings visualization 2000. VIS 2000 (Cat. No.00CH37145)

Desbrun M, Meyer M, Schröder P, Barr A (2001) Implicit fairing of irregular meshes using diffusion and curvature flow. SIGGRAPH

Digne J, de Franchis C (2017) The bilateral filter for point clouds. Image Process Line 7:278–287. https://doi.org/10.5201/ipol.2017.179

Durand F, Dorsey J (2002) Fast bilateral filtering for the display of high-dynamic-range images. ACM Trans Graph 21(3):257–266

Fleishman S, Drori I, Cohen-Or D (July, 2003) Bilateral mesh denoising. ACM Trans Graph 22(3):950–953. ISSN: 0730-0301. Association for Computing Machinery, New York, NY, USA. https://doi.org/10.1145/882262.882368

Hampel FR, Ronchetti EM, Rousseeuw PJ, Stahel WA (2005) Robust statistics: the approach based on influence functions. Wiley

Hildebrandt K, Polthier K (2004) Anisotropic filtering of non linear surface features. Comput Graph Forum. ISSN: 1467-8659. https://doi.org/10.1111/j.1467-8659.2004.00770.x

Hoppe H, DeRose T, Duchamp T, McDonald J, Stuetzle W (1992) Surface reconstruction from unorganized points. SIGGRAPH Comput Graph

Huber PJ (1981)Robust statistics. Wiley

Jones TR, Durand F, Desbrun M (2003) Non-iterative, feature-preserving mesh smoothing. ACM Trans Graph 22(3):943–949

Lange C, Polthier K (2005) Anisotropic smoothing of point sets. Comput Aided Geom Des 22(7):680–692

Li J (2009) Feature-preserving denoising of point-sampled surfaces. In: Proceedings of the 3rd WSEAS international conference on computer engineering and applications

Mattei E, Castrodad A (2016) Point cloud denoising via moving RPCA: Mrpca. *Computer Graphics Forum*, 2016

Mrázek, P, Weickert J, Bruhn A (2006) On robust estimation and smoothing with spatial and tonal kernels. Springer, Berlin, , pp 335–352

Ohtake Y, Belyaev AG, Seidel H-P (2002) Mesh smoothing by adaptive and anisotropic gaussian filter applied to mesh normals. In: In vision modeling and visualization, Eurographics Association. https://www.semanticscholar.org/paper/Mesh-Smoothing-by-Adaptive-and-Anisotropic-Gaussian-Ohtake-Belyaev/19b431c843f4b37d2218e7efcd8f64b6ff589c1f

Öztireli X, Guennebaud G, Gross M (2009) Feature preserving point set surfaces based on non-linear kernel regression. Comput Graph Forum

Park MK, Lee SJ, Jang Y, Lee YY, Lee KH (2013) Feature-aware filtering for point-set surface denoising. Comput Graph

Perona P, Malik J (1990) Scale-space and edge detection using anisotropic diffusion. IEEE Trans Pattern Anal Mach Intell 12(7):629–639. https://doi.org/10.1109/34.56205

Shen Y, Barner KE (2004) Fuzzy vector median-based surface smoothing. IEEE Trans Visual Comput Graph 10(3):252–265

Skrodzki M, Jansen J, Polthier K (2018) Directional density measure to intrinsically estimate and counteract non-uniformity in point clouds. Comput Aided Geom Des 64:73–89

Sun X, Rosin PL, Martin R, Langbein F (2007) Fast and effective feature-preserving mesh denoising. IEEE Trans Visual Comput Graph 13(5):925–938

Sun Y, Chen H, Qin J, Li H, Wei M, Zong H (2019) Reliable rolling-guided point normal filtering for surface texture removal. In: Computer graphics forum, vol 38 , issue no 7, pp 721–732. Wiley Online Library

Tasdizen T, Whitaker R, Burchard P, Osher S (2002) Geometric surface smoothing via anisotropic diffusion of normals. In: IEEE visualization, 2002. VIS 2002, pp 125–132

Taubin G (1999) A signal processing approach to fair surface design. Comput Graph (1999) (Proceedings of Siggraph '95)

Taubin G (2001a) Geometric signal processing on polygonal meshes. Eurographics State of the Art Reports

Taubin G (2001b) Linear anisotropic mesh filtering. In: IBM research report RC22213(W0110-051), IBM T.J. Watson Research

Tomasi C, R. Manduchi. Bilateral filtering for gray and color images. In: Iccv, vol 98, issue no 1

Trahanias PE, Venetsanopoulos AN (1993) Vector directional filters-a new class of multichannel image processing filters. IEEE Trans Image Process 2(4):528–534

Vallet B, Levy B (2008) Spectral geometry processing with manifold harmonics. Comput Graph Forum 27(2):251–260. https://doi.org/10.1111/j.1467-8659.2008.01122.x

Winkler G, Aurich V, Hahn K, Martin A, Rodenacker K (1998) Noise reduction in images: some recent edge-preserving methods. 138, sfb386. https://epub.ub.uni-muenchen.de/1527/. http://nbn-resolving.de/urn/resolver.pl?urn=nbn:de:bvb:19-epub-1527-2

Yadav SK, Reitebuch U, Skrodzki M, Zimmermann E, Polthier K (2018a) Constraint-based point set denoising using normal voting tensor and restricted quadratic error metrics. Comput Graph 74:234–243. ISSN: 0097-8493. https://doi.org/10.1016/j.cag.2018.05.014. https://www.sciencedirect.com/science/article/pii/S0097849318300797

Yadav SK, Kadas EM, Motamedi S, Polthier K, Hausser F, Gawlik K, Paul F, Brandt A (2018b) Optic nerve head three-dimensional shape analysis. J Biomed Opt 23(10):1–13

Yadav SK, Reitebuch U, Polthier K (2018c) Mesh denoising based on normal voting tensor and binary optimization. IEEE Trans Visual Comput Graph 24(8):2366–2379

Yadav SK, Reitebuch U, Polthier K (2019) Robust and high fidelity mesh denoising. IEEE Trans Visua Comput Graph 25(6):2304–2310

Yagou H, Ohtake Y, Belyaev AG (2003) Mesh denoising via iterative alpha-trimming and nonlinear diffusion of normals with automatic thresholding. Proce Comput Grap Int 2003:28–33

Yagou H, Ohtake Y, Belyaev A (2002) Mesh smoothing via mean and median filtering applied to face normals. In: Proceedings of geometric modeling and processing. Theory and Applications. GMP 2002. pp 124–131

Zhang W, Deng B, Zhang J, Bouaziz S, Liu L (2015) Guided mesh normal filtering. Comput Graph Forum 34(7):23–34

Zheng Y, Fu H, Au OK, Tai C (2011) Bilateral normal filtering for mesh denoising. IEEE Trans Visual Comput Graph 17(10):1521–1530

Zheng Y, Li G, Wu S, Y. Liu, and Y. Gao. Guided point cloud denoising via sharp feature skeletons. Vis Comput (2017)

Zheng Y, Li G, Xu X, Wu S, Nie Y (2018) Rolling normal filtering for point clouds. Comput Aided Geom Des 62:16–28. ISSN (0167-8396). https://doi.org/10.1016/j.cagd.2018.03.004. https://www.sciencedirect.com/science/article/pii/S0167839618300189

Unique Continuation on a Sphere for Helmholtz Equation and Its Numerical Treatments

Yu Chen and Jin Cheng

1 Introduction

Unique continuation means that if the solution of a partial differential equation vanishes on a "small" domain, it must vanish on the whole connected domain. The unique continuation properties of elliptic equations witness increasing attentions, and there have been many remarkable results, with the focus ranging from conditions on coefficients to requirements on data set (see e.g., (Daniel 2007; Isakov 2001; Vessella 2007; Saut and Scheurer 2017; Alexander Logunov and Eugenia Malinnikova 2018)). Several new aspects on unique continuation problem have also been considered, one of which is called the *partial unique continuation*. In some cases, the measured data for unique continuation can be on an analytic submanifold, e.g., on an analytic curve for harmonic functions (Cheng and Yamamoto 1998; Cheng et al. 1998) and wave equations (Cheng et al. 2005, 2002), along hypersurfaces for wave equations (Cheng et al. 1999) and second-order anisotropic hyperbolic systems (Cheng et al., 2005). In other cases, for some strong coupled partial differential systems, the known data is only available for part of the components of the solutions, which leads to the associated unique continuation problems (Wang et al. 2017).

There are a lot of applications of the unique continuation property. It provides ways to make use of the partial information of the solution (local information), to determine the other information of the solution (global information). These include recovering an unknown boundary from partial measurement (Aparicio and Pidcock 1996; Bukhgeim etal. 1998; Isakov 1993), identification of unknown time-varying boundaries (Vessella 2007), and obtaining global field from local interior measurement.

Y. Chen (✉)
School of Mathematics, Shanghai University of Finance and Economics, Shanghai 200433, China
e-mail: yu_chen11@fudan.edu.cn

J. Cheng
School of Mathematical Sciences and SKLCAM, Fudan University, Shanghai 200433, China

J. Cheng et al. (eds.), *Proceedings of the Forum "Math-for-Industry" 2018*,
Mathematics for Industry 35, https://doi.org/10.1007/978-981-16-5576-0_7

Relatively less investigated is the numerical realization of the unique continuation problem. In (Shuai et al. 2012) the numerical algorithms for line unique continuation of Helmholtz equation is provided based on the constructive proof of line unique continuation. The domain unique continuation is numerically considered in (Burman et al. 2018) based on optimal control using stabilized finite element method. The ill-posedness and the numerical errors would significantly affect the performance and accuracy of the numerical computation and give rise to difficulties in it. The stability does not hold even for external extension problem of analytic functions. Therefore, the conditional stability is crucial, with which we can construct stable solutions and reliable algorithms by, say, Tikhonov regularization (Cheng and Yamamoto 2000).

In this paper, we focus on the numerical unique continuation of elliptic equations on a sphere. The Helmholtz equation is mainly considered, which can be applied to inverse scattering problems. We will obtain the conditional stability and provide the numerical algorithm that generates convergent solutions, together with the numerical examples to illustrate the method. One direct application of practical significance is that the scattering field on the whole sphere can be recovered by partial measurements, which can be further utilized to determine the scatterer.

2 Main Results

Recall that in an inverse scattering problem, we determine the object or medium from the measurement outside, say, the far field pattern on a sphere. In a more practical situation, the measured data may be given only in an open set of the sphere. Then, it would be helpful if the field on the whole sphere could be recovered (see Fig. 1).

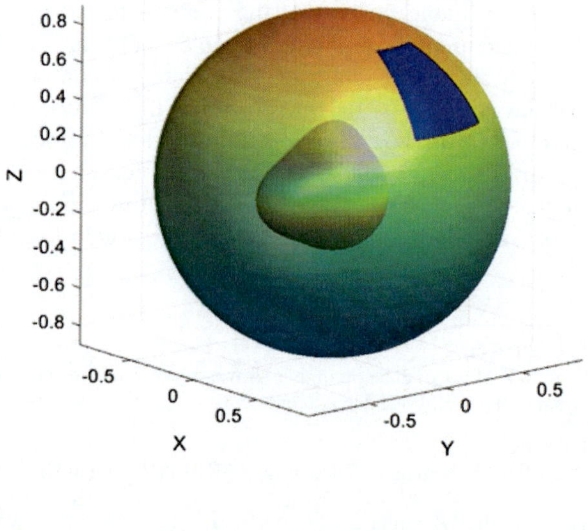

Fig. 1 Sketch of unique continuation with data on part of the outside sphere (the dark patch) in an obstacle scattering problem

Suppose that $S_R = \{x \in \mathbb{R}^3 \big| \, |x| = R\}$ is the sphere and Γ is an open set on S_R. We denote Ω a bounded domain with smooth boundary and assume that $S_R \subset \Omega \subset \mathbb{R}^3$. Then, the extension problem is to obtain u on S_R with

$$\Delta u + k^2 u = 0, \quad \text{in} \quad \Omega, \tag{1}$$

$$u = f, \quad \text{on} \quad \Gamma. \tag{2}$$

The main result on the conditional stability is stated in the following.

Theorem 1 *[Conditional stability] Suppose that $g = u \mid_{\partial\Omega}$, which satisfies*

$$\|g\|_{L^2(\partial\Omega)} \leq M, \tag{3}$$

then there exist constants $C(M, \Omega, \Gamma)$ and $\alpha \in (0, 1)$, which are independent of u, such that

$$\|u|_{S_R}\|_{C(S_R)} \leq C(M, \Omega, \Gamma)\|f\|^{\alpha}_{C(\Gamma)}. \tag{4}$$

As consequences of the above estimate, we have the following corollaries on unique continuation.

Corollary 1 *[Unique continuation in a simply connected domain] Suppose that Ω is a simply connected domain and $B_R \subset \Omega$ where $B_R = \{x \in \mathbb{R}^3 \big| \, |x| \leq R\}$. If $f = 0$, then $u = 0$ in Ω.*

Corollary 2 *[Unique continuation on a sphere] Suppose that Ω is not a simply connected domain such that $B_R \bigcap \Omega \neq B_R$. If $f = 0$, then we have $u = 0$ on S_R, but we may have $u \neq 0$ in Ω.*

Note that in the present case, the values of solution are only given on an open set of the sphere, i.e., a two dimensional surface rather than an open set in $\mathbb{R}^3$. Unlike the Cauchy problems for elliptic equations, no Neumann derivatives are given.

Proof of Theorem 1. Without loss of the generality, we assume that k^2 is not the eigenvalue of Laplace operator in Ω. The solution $u(x)$ can be represented by the single layer potential as

$$u(x) = \int_{\partial\Omega} \Phi(x, y)\mu(y)\mathrm{d}S(y), \quad x \in \Omega \quad \text{where} \quad \Phi(x, y) = \frac{e^{ik|x-y|}}{|x-y|},$$

where $\mu(y)$ is called the single layer potential density function.

By using spherical coordinates

$$(x, y, z) = (r \sin\theta \cos\phi, r \sin\theta \sin\phi, r \cos\theta)$$

and

$$(r, \theta, \phi) = (\sqrt{x^2 + y^2 + z^2}, \arctan \frac{\sqrt{x^2 + y^2}}{z}, \arctan \frac{y}{x}),$$

we have the representation of u on S_R,

$$u(R, \theta, \phi) = \int_{\partial\Omega} \Phi(x(R, \theta, \phi), y)\mu(y)\mathrm{d}S(y).$$

Without loss of generality, we assume $\Gamma = \{(x, y, z)(\theta, \phi)|0 < a_1 < \theta < a_2 < \pi, 0 < b_1 < \phi < b_2 < 2\pi\}$. The problem thus becomes: from $u(R, \theta, \phi)$, $(\theta, \phi) \in (a_1, a_2) \times (b_1, b_2)$ to find $u(R, \theta, \phi)$, $(\theta, \phi) \in (0, \pi) \times (0, 2\pi)$.

Let $z_1 = \theta_1 + i\theta_2$, $z_2 = \phi_1 + i\phi_2 \in \mathbb{C}$. One can construct an analytic function of two complex variables as

$$W(z_1, z_2) = \int_{\partial\Omega} \Phi((R, z_1, z_2), y)\mu(y)\mathrm{d}S(y). \tag{5}$$

It is easy to verify that there exists a constant δ, which depends on $dist(S_R, \partial\Omega)$, such that $W(z_1, z_2)$ is analytic in $U_1 \times U_2$ in $\mathbb{C}^2$, where

$$U_1 = \{z_1| -\delta < \Re z_1 < 2\pi + \delta; -\delta < \Im z_1 < \delta\}, \tag{6}$$
$$U_2 = \{z_2| -\delta < \Re z_2 < \pi + \delta; -\delta < \Im z_2 < \delta\}. \tag{7}$$

$\Re\zeta$ and $\Im\zeta$ denote the real and imaginary parts of $\zeta \in \mathbb{C}$, respectively. From the integral representation formula (5), it can be verified that

$$W(\theta, \phi) = u(R, \theta, \phi), \quad \text{for} \quad \theta, \phi \in \mathbb{R}.$$

Denote the segments

$$l_1 = \{z_1|a_1 \le \Re z_1 \le b_1; \quad \Im z_1 = 0\} \subset U_1,$$
$$l_2 = \{z_2|a_2 \le \Re z_2 \le b_2; \quad \Im z_2 = 0\} \subset U_2,$$

and consider the domain $V_1 \times V_2$ in $\mathbb{C}^2$ where $V_1 = U_1 \backslash l_1$, $V_2 = U_2 \backslash l_2$, the characteristic boundary of $V_1 \times V_2$ is

$$(\partial U_1 \cup l_1) \times (\partial U_2 \cup l_2).$$

By maximum principle, the maximum for the absolute value of an analytic function can only be reached on the characteristic boundary unless it is a constant.

Definition 1 $\mu_j(\zeta)$, $(j = 1, 2)$ is called the harmonic measure for U_j and l_j if it satisfies

$$\Delta\mu_j(\zeta) = 0, \quad \zeta \in U_j \backslash l_j \tag{8}$$
$$\mu_j(\zeta) = 0, \quad \zeta \in \partial U_j \tag{9}$$
$$\mu_j(\zeta) = 1, \quad \zeta \in l_j \tag{10}$$

One has the following properties of the harmonic measure, for details of which we refer to, e.g., (Friedman and Vogelius 1989; Kellogg 1953).

Lemma 1 *There exists a unique solution μ_j to (8)-(10), and $0 < \mu_j < 1$, for $z_j \in V_j$. For every point z^* in V_j, there exists a real harmonic function v_j, which is defined in the neighborhood of z^*, such that $\mu_j + iv_j$ is holomorphic in this neighborhood domain.*

According to (5), it can be proved by standard technique in PDE that

$$|W(z_1, z_2)| \leq C(\partial\Omega, S_R)M \equiv M_1, \quad (z_1, z_2) \in U_1 \times U_2. \tag{11}$$

On $l_1 \times l_2$,

$$|W(z_1, z_2)| = |u(R, \theta, \phi)| = |f|, \quad \text{for} \quad z_1 = \theta;\ z_2 = \phi.$$

Denote

$$\varepsilon = \max_{z_1 \in l_1; z_2 \in l_2} |W(z_1, z_2)| = \|f\|_{C(\Gamma)}. \tag{12}$$

Consider the function

$$U(z_1, z_2) = W(z_1, z_2)\exp\{(1 - \mu_1(z_1)\mu_2(z_2)) \ln\varepsilon + \mu_1(z_1)\mu_2(z_2) \ln M_1\}.$$

We now prove that

$$|U(z_1, z_2)| \leqslant |U(z_1, z_2)|_{C(\partial V_1 \times \partial V_2)} = M_1\varepsilon. \tag{13}$$

In fact, the equality in (13) is readily seen by noticing that $\max_{(z_1,z_2)\in\partial U_1\times\partial U_2} |U| = M_1$, $\max_{(z_1,z_2)\in l_1\times l_2} |U| = \varepsilon$ and the maximum principle of a holomorphic function. The inequality can be proved by assuming that there exists $\zeta' = (\zeta_1, \zeta_2) \in U_1 \times U_2 \backslash l_1 \times l_2$ such that

$$|U(\zeta')| = \max_{\zeta\in\overline{U}_1\times\overline{U}_2} |U| > \max_{\zeta\in\partial V_1\times\partial V_2} |U| = M_1\varepsilon. \tag{14}$$

Fix ζ_1 and denote $\mu_1' = \mu_1(\zeta_1)$. Let $\omega \subset V_2$ be a simply connected domain which contains ζ_2. For the harmonic $\mu_2(z_2)$ in ω, there exists a holomorphic function $\Psi(z_2)$ in ω such that $\Psi(z_2) := \mu_2(z_2) + iv_2(z_2)$ due to Lemma 1. Let

$$V(z_2) = W(\zeta_1, z_2)\exp\{(1 - \mu_1'\Psi(z_2)) \ln\varepsilon + \mu_1'\Psi(z_2) \ln M_1\}.$$

It is readily seen that $V(z_2)$ is a holomorphic function in ω which reaches maximum at the interior point ζ_2. Therefore, the maximum principle for a holomorphic function implies

$$|V(z_2)| = \text{constant}, \quad z_2 \in \omega.$$

Subsequently,

$$|U(\zeta_1, z_2)| = \text{constant}, \quad z_2 \in \omega.$$

By repeating the above steps for $\zeta_2' \in \omega$ that differs from ζ_2 and considering that $\Psi(z_2) \in C(\bar{U}_2)$, we can extend the constant value area until we have

$$|U(\zeta_1, z_2)| = |U(\zeta_1, \zeta_2)| = \text{constant}, \quad z_2 \in U_2.$$

Similarly, we have

$$|U(z_1, z_2)| = |U(\zeta_1, z_2)| = |U(\zeta_1, \zeta_2)|, \quad (z_1, z_2) \in U_1 \times U_2,$$

which contradicts (14) and (13) is true.

Therefore, we arrive at the stability result

$$|W(z_1, z_2)| \leq M_1^{1-\mu_1(z_1)\mu_2(z_2)} \varepsilon^{\mu_1(z_1)\mu_2(z_2)}.$$

Since on the real axis the harmonic measure $\mu_1(x) \geq \mu_1(2\pi) > 0, \forall x \in [b_1, 2\pi]$ and $\mu_1(x) \geq \mu_1(0) > 0, \forall x \in [0, a_1]$, there exists a constant $\alpha_1 \in (0, 1)$ depending on U_1 and l_1 such that $\mu_1(x) > \alpha_1, \forall x \in l_1$. Similarly, there exists a constant $\alpha_2 \in (0, 1)$ such that $\mu_2(x) > \alpha_2, \forall x \in l_2$. Denote $\alpha = \alpha_1\alpha_2$, we finally have

$$\|u|_{S_R}\|_{C(S_R)} \leq C(M, \Omega, \Gamma)\|f\|_{C(\Gamma)}^{\alpha}. \tag{15}$$

Corollary 3 *Suppose that $g = u\ |_{\partial\Omega}$, which satisfies*

$$\|g\|_{L^2(\partial\Omega)} \leq M, \tag{16}$$

then there exist constants $C(M, \Omega, \Gamma)$ and $\alpha \in (0, 1)$, which are independent of u, such that

$$\|u|_{S_R}\|_{L^2(S_R)} \leq C(M, \Omega, \Gamma)\|f\|_{L^2(\Gamma)}^{\alpha}. \tag{17}$$

Proof According to Sobolev embedding theorem and interpolation theorem (Adams Robert et al. 2003; Hebey Emmanuel 2000), we have

$$\|f\|_{C(\Gamma)} \leq C_1\|f\|_{H^2(\Gamma)} \leq C_2\|f\|_{L^2(\Gamma)}^{1/2}\|f\|_{H^4(\Gamma)}^{1/2}. \tag{18}$$

Since $\overline{B}_R \subset \Omega\ (\partial B_R = S_R)$, it can be seen from single layer potential representation that $\|f\|_{H^4(\Gamma)}^{1/2} \leq C_3(M, \partial\Omega, \Gamma)$. Therefore,

$$\|f\|_{C(\Gamma)} \leq C_4(M, \Omega, \Gamma)\|f\|^{1/2}_{L^2(\Gamma)}. \tag{19}$$

Considering that $\|u|_{S_R}\|_{L^2(S_R)} \leq C_3(S_R)\|u|_{S_R}\|_{C(S_R)}$ and combining (4), we arrive at (17).

3 Numerical Method and Examples

The unique continuation can be numerically realized by recovering the single-layer potential density function, that is, to solve the integral equation with analytical integral kernel. We consider the minimum norm solution in finite dimensional test spaces. Denote the test space on $\partial\Omega$ as $V_n = span\{v_j\}_{j=1}^n$. Here we can choose v_j the standard basis for spherical polynomials based on the Legendre polynomials and associated Legendre functions. It holds that $V_n \subset V_{n+1} \subset H^1(\partial\Omega)$ and $\bigcup_{n=1}^{\infty} V_n$ is dense in $L^2(\partial\Omega)$.

Algorithm

We choose $\{x_1, x_2, ..., x_m\} \in \Gamma$ and construct the matrix A as

$$A = [A_{ij}], \quad A_{ij} = \int_{\partial\Omega} \Phi(x_i, y)v_j(y)\mathrm{d}S(y), \quad (i = 1, 2, ..., m;\ j = 1, 2, ..., n).$$

Find the minimum norm solution $\tilde{\xi}$ to the equation $A\xi = b$, where $b = (f(x_1), f(x_2), ..., f(x_m))^T$. The approximation solution can then be constructed as

$$\tilde{u}(x) = \Psi(x) \cdot \tilde{\xi},$$

where

$$\Psi(x) = \left(\int_{\partial\Omega} \Phi(x, y)v_1(y)\mathrm{d}S(y), ..., \int_{\partial\Omega} \Phi(x, y)v_n(y)\mathrm{d}S(y)\right).$$

In practical implementation, a singular value tolerance may need to be introduced in computation of Moore–Penrose pseudoinverse to ensure the boundness of the solution. Then, singular values of A that are smaller than the tolerance will be treated as zero.

We interpret the above discrete form in terms of interpolation spaces to facilitate the error estimate. Denote $K\mu = \int_{\partial\Omega} \Phi(x, y)\mu(y)\mathrm{d}S(y)$ with K being the integral operator. The above discrete form $A\xi = b$ can be viewed in the interpolation space as $\hat{K}\mu_n = \hat{f}$, where $\mu_n(y) = \sum_{j=1}^n \tilde{\xi}^j v_j(y)$,

$$\hat{K}\mu_n = \sum_{i=1}^{m} \varphi_i(x) \int_{\partial\Omega} \left(\Phi(x_i, y) \sum_{j=1}^{n} \tilde{\xi}^j v_j(y) \right) \mathrm{d}S(y), \quad \hat{f} = \sum_{i=1}^{m} \varphi_i(x) f(x_i).$$

$\varphi_i(x)$ are taken as the basis of piecewise constant function space. The class of piecewise constant functions on Γ is dense in $L^2(\Gamma)$. The approximation solution can be written in operator form as $\Psi(x) \cdot \tilde{\xi} = K\mu_n$. Let μ_0 be the exact solution to $K\mu_0 = f$, which also indicates $\hat{K}\mu_0 = \hat{f}$.

Based on Corollary 3 and the denseness of test spaces for μ, we have the following error estimate for the solution $\tilde{u} = K\mu_n$.

Theorem 2 *[Convergence] Suppose that* $\|\mu_0\|_{L^2(\partial\Omega)} \leq M$, *there exist constants* C, $\alpha \in (0, 1)$ *and* $\epsilon(n)$ *such that,*

$$\|\tilde{u}(x) - u(x)\|_{L^2(S_R)} \leq C(\Omega, S_R, M)(\epsilon(n))^\alpha, \tag{20}$$

where $\lim_{n\to\infty} \epsilon(n) = 0$.

Proof For the residual we have

$$\|K\mu_n - f\|_{L^2(\Gamma)} \leq \|K\mu_n - \hat{K}\mu_n\|_{L^2(\Gamma)} + \|\hat{K}\mu_n - \hat{f}\|_{L^2(\Gamma)} + \|\hat{f} - f\|_{L^2(\Gamma)}$$

where $\|K\mu_n - \hat{K}\mu_n\|_{L^2(\Gamma)}$ and $\|\hat{f} - f\|_{L^2(\Gamma)}$ converge as m increases due to the denseness of the interpolation space on Γ. The remaining term can be estimated as

$$\begin{aligned} \|\hat{K}\mu_n - \hat{f}\|_{L^2(\Gamma)} &= \inf_{\mu \in V_n} \|\hat{K}\mu - \hat{f}\|_{L^2(\Gamma)} \leq \|\hat{K}P_n\mu_0 - \hat{f}\|_{L^2(\Gamma)} \\ &\leq \|\hat{K}P_n\mu_0 - \hat{K}\mu_0\|_{L^2(\Gamma)} + \|\hat{K}\mu_0 - \hat{f}\|_{L^2(\Gamma)}. \end{aligned}$$

Due to the denseness of the test space for V_n (the standard spherical polynomial space here) in $L^2(\partial\Omega)$, we have $\|P_n\mu_0 - \mu_0\|_{\partial\Omega} \leq C_1(M, \Omega)\epsilon(n)$, where $\lim_{n\to\infty} \epsilon(n) = 0$. Therefore,

$$\|\hat{K}P_n\mu_0 - \hat{K}\mu_0\|_{L^2(\Gamma)} \leq \|\hat{K}\|\|P_n\mu_0 - \mu_0\|_{L^2(\partial\Omega)} \leq C_2(S_R, \Omega, M)\epsilon(n),$$

and $\|\tilde{u} - f\|_{L^2(\Gamma)} \leq C_3(S_R, \Omega, M)\epsilon(n)$.

Moreover, since μ_n is the minimum norm solution and a singular value tolerance constant is introduced, μ_n is bounded and depends on K which further depends on S_R and Ω. Then, $\|\tilde{u}\|_{L^2(\partial\Omega)} = \|(K\mu_n)|_{\partial\Omega}\|_{L^2(\partial\Omega)} \leq C_4(S_R, \Omega, M)$ and the conditional stability of unique continuation can therefore be applied. According to (4) and the residual on Γ, we arrive at the convergence (20) on S_R.

Example 1 We now give numerical results to demonstrate the applicability of the proposed method. The first example considers the Helmholtz equation with exact solution $u = \cos(\pi x)\cos(\pi y)\cos(\pi z)$ $(k = \sqrt{3}\pi)$ and the radius of the sphere $R = 0.9$. The data on the uniformly spaced 25×25 $(m = 25^2)$ grid on the patch $(\theta, \phi) \in$

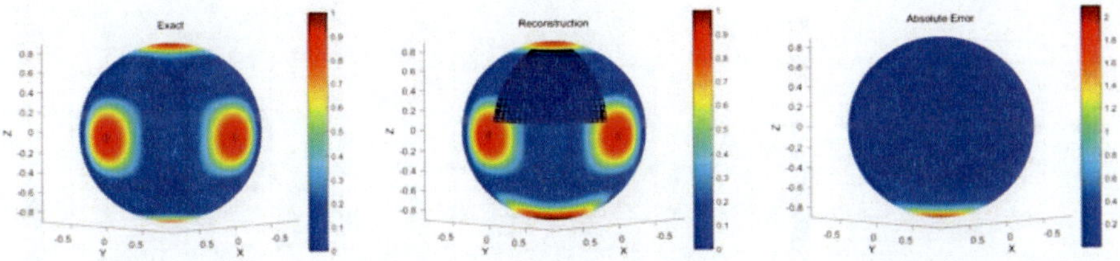

(a) Exact solution (b) Reconstructed solution (c) Absolute error

Fig. 2 Unique continuation of Helmholtz equation with exact solution

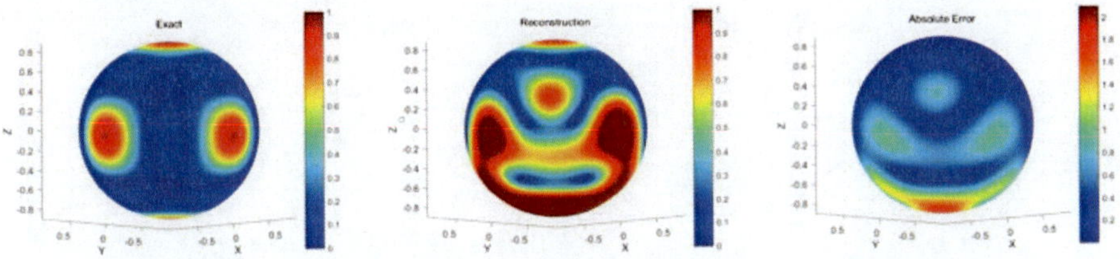

(a) Exact solution (b) Reconstructed solution (c) Absolute error

Fig. 3 Unique continuation of Helmholtz equation with exact solution (back view)

$[0.1\pi, 0.45\pi] \times [0, 0.45\pi]$ was assumed to be known. On the auxiliary boundary for μ we use the sphere harmonic basis with order $\hat{n} = 20$ $(n = (\hat{n} + 1)^2)$.

The result is displayed in Fig. 2. Figure 2b gives the reconstructed solution where the data is given on the black patch. The result on the other semisphere is shown in Fig. 3. From the absolute error shown in Fig. 2c, we can see that the reconstructed result coincides with the exact solution well at least in the front hemisphere which contains the measurement patch, with the absolute error being less than 0.1. The error on the other semisphere is large. This is because the stability is of Hölder-type, and from the proof of Theorem 1, we can see that the exponent α depends on the harmonic measure which becomes lower when getting further from the data area.

Example 2 The second example is the unique continuation of the far field of sound-soft peanut scattering of one incident plane wave in homogeneous medium, with wave number $k = \sqrt{3}\pi$ and direction $\xi = (-1/\sqrt{2}, 0, -1/\sqrt{2})$. The Sommerfeld radiation condition is imposed. The scattered field $u^s(x)$ corresponding to the incoming wave $u^i(x) \doteq e^{ikx\cdot\xi}$ satisfies

$$u(x) = e^{ikx\cdot\xi} + u^s(x).$$

The far field pattern u_∞ is defined as the first coefficient in the asymptotic expansion of u^s:

$$u^s(x) = \frac{e^{ik|x|}}{|x|} u_\infty\left(\frac{x}{|x|}\right) + O\left(\frac{1}{|x|^2}\right).$$

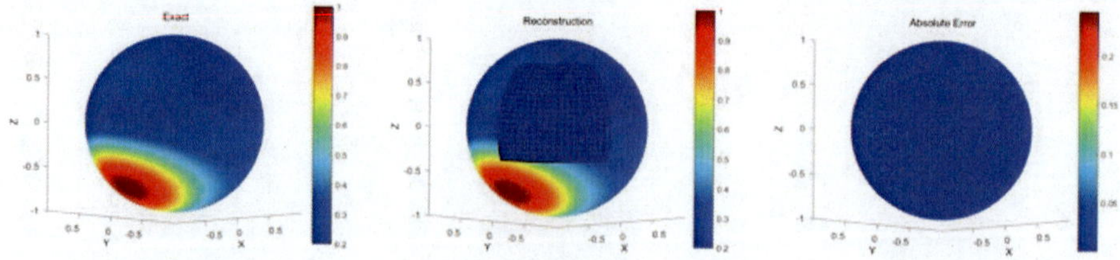

(a) Exact far field (b) Reconstructed far field (c) Absolute error

Fig. 4 Unique continuation of far field of a sound-soft scattering problem

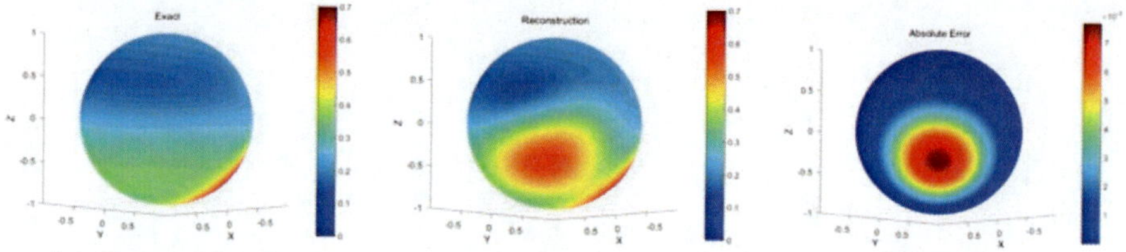

(a) Exact far field (b) Reconstructed far field (c) Absolute error

Fig. 5 Unique continuation of far field of a sound-soft scattering problem (back view)

The sound-soft obstacle is axial symmetry and is described through its radial distance in terms of the polar angle θ by

$$r(\theta) = 0.5\left(\cos^2\theta + 0.3\sin^2\theta\right)^{1/2}.$$

The direct scattering field and the corresponding far field are obtained by the Galerkin's method, for details of which we refer to (Atkinson 1982; Lin 1985; Colton David L and Kress Rainer 1998).

The data on the uniformly spaced 18×18 ($m = 18^2$) grid on the patch $(\theta, \phi) \in [0.2\pi, 0.6\pi] \times [1.0\pi, 1.5\pi]$ was assumed to be known. On the auxiliary boundary for μ, we use the sphere harmonic basis with order $\hat{n} = 10$, i.e., $n = (\hat{n} + 1)^2$.

Figure 4b is the reconstructed solution where the data on the black patch is known. Since the solution is complex valued, we give the distribution of its norm $|u^s|$ on the sphere. The result on the opposite hemisphere is given in Fig. 5. From the absolute error distribution shown in Figs. 4c and 5c, we can see that the reconstructed result coincides with the true value well in the front hemisphere, the error on the back hemisphere is relatively large which can also be explained by the stability of the problem.

Remark 1. Note that though the scattering field is numerically obtained, it is calculated through single or double layer potentials and therefore satisfies Helmholtz equation. Accordingly, the unique continuation property with the numerical method are still valid.

Remark 2. Here we carry out the reconstruction of far field pattern and have not considered the inverse scattering problem.

The above examples illustrate the applicability of the numerical method. We note that the reconstruction error relies on the choices of the measurement area and the discretization on the auxiliary surface. The details of the performance regarding this method will be discussed in our forthcoming paper.

4 Conclusion

Unique continuation provides ways to infer information from local to global. The conditional stability of unique continuation on a sphere obtained in this paper makes it possible to design reliable numerical algorithms and estimate errors. The present method can also be extended to unique continuation problems of other elliptic equations. The main results together with the corresponding numerical method show potentials in applications in inverse problems and optimal design problems. One such topic of interest is the inverse scattering problem with partial far field measurement.

Acknowledgements The authors thank the referee for useful comments. This work was supported by the National Science Foundation of China (No. 11971121).

References

Adams RA, Fournier JJF (2003) Sobolev spaces, 2nd edn. Elsevier

Alessandrini G, Rondi L, Rosset E et al (2009) The stability for the Cauchy problem for elliptic equations. Inverse Prob 25(12):1541–1548

Logunov A, Eugenia M (2018) Quantitative propagation of smallness for solutions of elliptic equations. In: Proceedings of inernational congress of mathematicians—2018, Rio de Janeiro, vol 2, pp 2357-2378

Aparicio ND, Pidcock MK (1996) The boundary inverse problem for the Laplace equation in two dimensions. Inverse Probl 12(5):565

Atkinson KE (1982) The numerical solution of Laplace's equation in three dimensions. SIAM J Numer Anal 19(2):263–274

Bukhgeim AL, Cheng J, Yamamoto M (1998) On a sharp estimate in a non-destructive testing: determination of unknown boundaries Preprint 98-3 Graduate School of Mathematical Sciences, The University of Tokyo

Bukhgeim AL, Cheng J, Yamamoto M (1999) Stability for an inverse boundary problem of determining a part of a boundary. Inverse Prob 15(4):1021

Burman E, Hansbo P, Larson M (2018) Solving ill-posed control problems by stabilized finite element methods: an alternative to Tikhonov regularization. Inverse Prob 34(3):035004

Cheng J, Hon YC, Yamamoto M (1998) Stability in line unique continuation of harmonic functions: general dimensions. J Inverse Ill-Posed Prob 6(4):319–326

Cheng J, Yamamoto M (1998) Unique continuation on a line for harmonic functions. Inverse Prob 14(4):869

Cheng J, Yamamoto M, Zhou Q (1999) Unique continuation on a hyperplane for wave equation. Chin Ann Math Ser B 20(4):385–392

Cheng J, Yamamoto M (2000) One new strategy for a priori choice of regularizing parameters in Tikhonov's regularization. Inverse Prob 16(4):L31–L38(8)
Cheng J, Yamamoto M (2000) The global uniqueness for determining two convection coefficients from Dirichlet to Neumann map in two dimensions. Inverse Prob 16(3):L25
Cheng J, Ding G, Yamamoto M (2002) Uniqueness along a line for an inverse wave source problem. Commun Partial Differ Equat 27(9–10):2055–2069
Cheng J, Hon YC, Yamamoto M (2004) Conditional stability for an inverse Neumann boundary problem. Appl Anal 83(1):49–62
Cheng J, Yamamoto M (2004) Determination of two convection coefficients from Dirichlet to Neumann map in the two-dimensional case. SIAM J Math Anal 35(6):1371–1393
Cheng J, Peng L, Masahiro Yamamoto (2005) The conditional stability in line unique continuation for a wave equation and an inverse wave source problem. Inverse Prob 21(21):1993
Cheng J, Lin CL, Nakamura G (2005) Unique continuation along curves and hypersurfaces for second order anisotropic hyperbolic systems with real analytic coefficients. Proc Am Math Soc 133(8):2359–2367
Colton DL, Kress R (1998) Inverse acoustic and electromagnetic scattering theory. In: Inverse acoustic and electromagnetic scattering theory, 2nd edn. Springer, Berlin
Deckelnick K, Günther A, Hinze M (2009) Finite element approximation of Dirichlet boundary control for elliptic PDEs on two- and three dimensional curved domains. SIAM J Control Optim 48(4):2798–2819
Friedman A, Vogelius M (1989) Determining cracks by boundary measurements. Indiana Univ Math J 38(3):527–556
Gilbarg D, Trudinger NS (1983) Elliptic partial differential equations of second order, 2nd edn. Springer, Berlin
Hebey E (2000) Nonlinear analysis on manifolds: sobolev spaces and inequalities. Am Math Soc
Hsiao GC, Wendland WL (1977) A finite element method for some integral equations of the first kind. J Math Anal Appl 58(3):449–481
Isakov V (1993) New stability results for soft obstacles in inverse scattering. Inverse Prob 9:535–43
Isakov V (2001) On the uniqueness of the continuation for a thermoelasticity system. SIAM J Math Anal 33(3):509–522
Isakov V (2006) Inverse problems for partial differential equations, 2nd edn. Springer, Berlin
Kellogg OD (1953) Foundations of potential theory. Dover Publications Inc, New York
Lin TC (1985) The numerical solution of Helmholtz's equation for the exterior Dirichlet problem in three dimensions. SIAM J Numer Anal 22(4):670–686
Lu S, Xu B, Xu X (2012) Unique continuation on a line for the Helmholtz equation. Appl Anal 91(9):1761–1771
Saut JC, Scheurer B (2017) Unique continuation for some evolution equations. J Differ Equat 66(66):118–139
Daniel T (2007) Unique continuation for solutions to PDE's; between Hormander's theorem and Holmgren' theorem. Commun Partial Differ Equat 20(5–6):855–884
Vessella S (2007) Quantitative estimates of unique continuation for parabolic equations, determination of unknown time-varying boundaries and optimal stability estimates. Inverse Prob 24(2):213–229
Wang Y, Liu Y, Cheng J (2017) A new unique continuation property for the Lamé system in two dimensions. Scientia Sinica (Math) 47(10):1327–1334

Huberization Image Restoration Model from Incomplete Multiplicative Noisy Data

Xiaoman Liu and Jijun Liu

1 Introduction

Image processing can be roughly divided into three different kinds, namely image deblurring, image enhancement, and image restoration, with the main purpose of obtaining the clearer image from its noisy measurement. For a bounded connected domain $\Omega \subset \mathbb{R}^2$ (a rectangle in general Aubert and Vese 1997), let $f(\boldsymbol{x})$ be the gray function of an image defined in Ω. In general, we are given $g^\delta(\boldsymbol{x})$, the noisy blurred picture of $f(\boldsymbol{x})$, due to the blurring process and the unavoidable errors in measurements. Such a blurring process can be modeled by

$$g^\delta(\boldsymbol{x}) = \int_{\mathbb{R}^2} k(\boldsymbol{x}, \boldsymbol{y}) f(\boldsymbol{y}) d\boldsymbol{y} + \xi := \mathbf{K}[f](\boldsymbol{x}) + \xi, \quad \boldsymbol{x} \in \Omega, \tag{1}$$

where we define $f(\boldsymbol{y}) \equiv 0$ in $\mathbb{R}^2 \setminus \overline{\Omega}$ for this blurring procedure with known point spread function (PSF) $k(\boldsymbol{x}, \boldsymbol{y})$, while $\xi \in \Omega$ is the additive noise describing the effect of random processes that occur in nature, such as white Gaussian noise (Zhu and Liu 2015; Hintermüller and Rincon-Camacho 2010; Guo et al. 2009; Clason et al. 2010).

The optimization scheme is one of the classical ways to the reconstruction of $f(\boldsymbol{x})$ based on the model (1) with additive noise, which minimizes the Tikhonov cost functional

$$J_0(f) = \frac{1}{2}\|\mathbf{K}[f] - g^\delta\|^2_{L^2(\Omega)} + \alpha \mathbf{L}[f] \tag{2}$$

with some penalty term $\mathbf{L}[f]$ and the regularization parameter $\alpha > 0$, where the operator $\mathbf{L}$ represents the priori regularity image f. In this work, we fix Gaussian blurring function $k(\boldsymbol{x}, \boldsymbol{y})$ with corresponding blurring operator $\mathbf{K}$ being the Dirac

X. Liu
Nanjing Agricultural University, Nanjing 210095, People's Republic of China

X. Liu (✉) · J. Liu
Southeast University, Nanjing 210096, People's Republic of China
e-mail: liuxm@njau.edu.cn

J. Cheng et al. (eds.), *Proceedings of the Forum "Math-for-Industry" 2018*,
Mathematics for Industry 35, https://doi.org/10.1007/978-981-16-5576-0_8

impulse function $\delta(\boldsymbol{x}-\boldsymbol{y})$, and it follows that $\mathbf{K}$ is an identity operator $\mathbf{I}$, i.e., considering the image restoration only from relative random noise contaminations without any blurring process.

By now, there already exist extensive work for image restoration using noisy measurement with absolute error. However, if the noisy level depends on the amplitude of the signal, which is represented by the relative error, the restoration becomes more complicated since the high frequencies of the images will be contaminated seriously. Let noise ξ be the random noise by multiplying σ times on the image, i.e.,

$$g^{\sigma}(\boldsymbol{x}) = f(\boldsymbol{x}) + \sigma \cdot f(\boldsymbol{x}), \quad \boldsymbol{x} \in \Omega, \tag{3}$$

which is called relative random noise or multiplicative noise. σ is relative noise level.

Generally, the given noisy measurement data of the image g^{σ} may be incomplete, leading to non-unique reconstruction of the desired image in principle. In this case, we can only find some approximate reconstruction in terms of the insufficient noisy data. Additionally, in many engineering configurations, instead of the spatial noisy data $g^{\sigma}_{m,n}$ for each pixel $\Omega_{m,n}$, the practical measurement data may be the incomplete frequency data, which are specified at finite number of discrete frequencies within some band-limited interval. For example, in magnetic resonance imaging (MRI), the data collected by an MR scanner are, roughly speaking, in the frequency domain (called K-space data) rather than in the spatial domain. For this stage, the model (2) should be replaced by

$$J_{\alpha}(f) = \frac{1}{2}\|\mathcal{P}\mathcal{F} \circ f - \mathcal{P} \circ \hat{g}^{\sigma}\|^2_{L^2(S)} + \alpha \mathbf{L}[f], \tag{4}$$

where $\mathcal{F}$ is the two-dimensional Fourier transform converting into frequency data $\hat{f}(\boldsymbol{\omega}) := \mathcal{F} \circ f(\boldsymbol{x}) \in \Omega' \subset \mathbb{C}^2, \mathcal{P} \circ \hat{g}^{\sigma} = \mathcal{P} \circ (\hat{f} + \hat{f} \cdot \sigma) \in S$ is the partial Fourier data with multiplicative noise, and $\mathcal{P}$ is the linear sampling operator, projecting the full frequency data into a lower-dimensional space $S \subset \Omega'$. Although it has been proven that for some special images, the image can be reconstructed exactly at almost probability 1 as the solution to some l^1 minimization problem using incomplete frequency data (Candès et al. 2006), the reconstruction for general images using band-limited frequency data is still very hard, for which we should keep some balance between the insufficiency of measurement data, denoising, and edge-preserving.

For different forms of penalty term $\mathbf{L}[f]$ such as $\|f\|_{l^1}$, $\|f\|_{l_{1-2}}$, $\|f\|_{l_q}$ $(q \in (0, 1))$ and $\|\nabla f\|_{l^2}$ (Hintermüller and Rincon-Camacho 2010; Guo et al. 2009; Clason et al. 2010; Tibshirani 1996; Yin et al. 2008), there have been thoroughly researches on image restoration based on minimizing (2), which denoise the additive absolute noise very well. In compressive sensing (CS) theory, the l^1 penalty term represents the sparsity of image. There are lots of researches about the evolution of l^1-norm in CS theory, like truncated norm denoted as $l_{t,1}$ and l_{1-2} (Ma et al. 2017). In (Fan et al. 2016), the authors proposed a CS recovery model by considering a non-convex smoothed function to approximate the rank, denoted as a low-rank regularization. On the other hand, the function f is often of some sharp jumps representing the edges of an image.

So it is natural to cooperate this *a-priori* requirement into the reconstruction model by also introducing the total variation (TV) penalty term $|f|_{TV} := \|\nabla f\|_{l^1}$ (Rudin et al. 1992). In the recent works, the norm-based regularization of the gradient of the image has many evolutions. Hintermüller and Wu expanded the scope of solvers for l_q-norm-based gradient of the image called TV^q-regularization (Hintermüller and Wu 2013). They proposed method considered a Huberization of the non-Lipschitz l_q-norm and combined a reweighting technique for handling the non-convexity with primal-dual semi-smooth Newton methods for image restoration (Chan et al. 1999; Hintermüller and Stadler 2003; Kunisch 2004; Hintermüller and Stadler 2006). Bredies introduced a novel concept of total generalized variation (TGV) of a function u and developed the mathematical theory in Bredies et al. (2010). It is equivalent to TV in terms of edge preservation and noise removal, while it can also be applied in imaging situations where the assumption that the image is piecewise constant is not valid. His group also proposed TGV can be applied for image denoising and deblurring on MRI image reconstruction (Knoll et al. 2011).

To this end, we define the general solution of (4) as a minimizer of the cost functional with two penalty terms: one is the standard TV penalty term ensuring the edge-preserving of the image, and the other is the wavelet transform penalty (Zhu and Shi 2013) guaranteeing the sparsity of the image based on the compressed sensing (CS) technique (Candès et al. 2006; Donoho 2006). More precisely, only considering the denoising process, i.e., $\mathbf{K} \equiv \mathbf{I}$ (identity operator), we construct several functionals

$$J(f) := \frac{1}{2}\|\mathcal{PF} \circ f - \mathcal{P} \circ \hat{g}^\sigma\|^2_{L^2(S)} + \alpha_1\|\Psi \circ f\|_{l^1} + \alpha_2|f|_{TV}, \tag{5}$$

where Ψ is wavelet operator. For a signal f expanded by $f = \langle f, \psi_{m,n}\rangle_{\mathbb{R}^{N\times N}}$ (Daubechies et al. 2004), we call f sparse under the orthogonal base $\{\psi_{m,n} : m, n = 1, \cdots, N\}$.

It is well known that both $\|\cdot\|_{l^1}$ and $|\cdot|_{TV}$ are not differentiable at zero point. To overcome this difficulty in numerical implementations, lots of the research add some small perturbation $\beta > 0$ which is introduced to make the absolute value function differentiable. Consider the l^1-norm of the gradient of the image with Charbonnier smooth approximation (Aubert and Kornprobst 2006), i.e., TV penalty term defined by $|f|_{TV,\beta} = \|\nabla f\|_{l^1,\beta} = \sum_{m,n=1}^N \phi_\beta(|\nabla f|)$ where

$$\phi_\beta(s) = \sqrt{s^2 + \beta}, \tag{6}$$

The method by adding the positive threshold β is called the Charbonnier smooth approximation with Charbonnier function (6). Even though the Charbonnier function has performed almost the same as the Huber function and better than the Green function approximation (Huber 1964), it has the worst theoretical approximation (Kalmoun 2018). Huber function is non-quadratic but convex which is defined as following

$$\phi_\epsilon(s) = \begin{cases} \frac{s^2}{2\epsilon}, & |s| \le \epsilon, \\ |s| - \frac{\epsilon}{2}, & |s| > \epsilon, \end{cases} \tag{7}$$

where $\epsilon > 0$ is a small parameter. It is used as a smooth approximation of l^1-norm in Madsen and Nielsen (1993), denoted as $\|f\|_{l^1,\epsilon} = \sum_{m,n=1}^{N} \phi_\epsilon(f)$. The most useful advantage by using Huber approximation is that it could smooth small-scale noise by the quadratic function part for arguments below a threshold ϵ, while preserving discontinuities at edge regions by the linear function part above the threshold. In this paper, we consider the unconstraint cost functional

$$J_{\alpha,\epsilon}(f) := \frac{1}{2}\|\mathcal{P}\mathcal{F} \circ f - \mathcal{P} \circ \hat{g}^\sigma\|_{L^2(S)}^2 + \alpha_1 \|\Psi \circ f\|_{l^1,\epsilon} + \alpha_2 |f|_{TV,\epsilon} \tag{8}$$

with $\alpha := (\alpha_1, \alpha_2) > 0$ as the penalty parameters, small perturbation $\epsilon > 0$ to make two penalty terms differentiable.

The paper is organized as follows. In Sect. 2, an efficient algorithm for the double regularizing optimizing model with Huberization l^1 penalty and TV penalty is proposed. In Sect. 3, some real magnetic resonance imaging (MRI) images are used to test in the numerical experiments, and it is to demonstrate effectiveness of our method in image restoration with high-level relative random noise from incomplete frequency data. Finally, the conclusion is given in Sect. 4.

2 The Algorithm for the Double Regularizing Image Restoration Model

Suppose $f \in \mathbb{R}^{N^2 \times 1}$ is a vector formed by stacking the columns of an $N \times N$ two-dimensional image array $f_{m,n} (m, n = 1, \cdots, N)$. In compressive sensing (CS), $\Psi \circ f$ is a vector meaning the sparse representation for image $f(\boldsymbol{x}_{m,n}) := f_{m,n} \in \mathbb{R}^{N \times N}$ with respect to orthogonal wavelet basis ψ, i.e., $(\Psi \circ f)_l := \langle f_{m,n}, \psi_{m,n}\rangle$ where $l := l(m, n) = (n-1)N + m, l = 1, \cdots, N^2$. To recover a sparse image signal, the bases $\psi_{m,n}$ can be constructed in terms of one-dimensional discrete wavelet transform (DWT), which constructs four two-dimensional function families and each function family is generated from the one-dimensional scaling function and mother wavelet function (Mallat 2009). In this work, we choose Danbechies wavelet bases as the wavelet operator. Let $K = \|\Psi \circ f\|_{l^0}$ be the number of nonzero elements in $\Psi \circ f$, R denotes $K \times N^2$ ($K \ll N^2$) measurement matrix such that $Rf = \hat{g}^\sigma$, where $\hat{g}^\sigma$ is an observed frequency data vector. Unfortunately, the l^0 decoder is generally NP-hard, so a common substitute for solving the minimizing $\|\cdot\|_0$ problem is the well-known basis pursuit problem (Chen et al. 2001), i.e., we use the l^1 norm instead of l^0. Then to recover f from observed frequency data, $\hat{g}^\sigma$ is equivalent to solve the following constrained optimization problem

$$\min_f \|\Psi \circ f\|_{l^1}, \quad s.t. \ Rf = \hat{g}^\sigma. \tag{9}$$

In compressed sensing (CS) theory, R is a partial Fourier matrix, i.e., $R = PF_t$, $P \in \mathbb{R}^{K \times N^2}$ is consisted of $K \ll N^2$ rows of the identity matrix, $F_t := F \otimes F$ is a two-dimensional discrete Fourier matrix with components

$$(F)_{m,n} = e^{-i\frac{2\pi}{N}n(m+\frac{N}{2})}, \quad m, n = 0, \cdots, N-1. \tag{10}$$

With the recent research (Liu and Liu 2019), the unconstrained version of problem (9) with adding TV regularizing term is

$$\min_f \left\{ \frac{1}{2} \|PF_t f - P\hat{g}^\sigma\|_{l^2}^2 + \alpha_1 \|\Psi \circ f\|_{l^1} + \alpha_2 |f|_{TV} \right\}, \tag{11}$$

where P is the sampling matrix consisting of only $K < N$ rows of the identity matrix I, $\alpha_1, \alpha_2 > 0$ are positive parameters that determine the trade-off among the fidelity term, i.e., the wavelet sparsity term and TV penalty term, and $\|\cdot\|_{l^2}$ denotes the Euclidean norm.

Since the second and the third terms in (11) are not Frechet differentiable at point $f = 0$, by the standard arguments (Liu and Zhu 2014; Zhu and Shi 2013), the approximation with Charbonnier function is used to differentiate. However, the used regularization with Charbonnier approximation is C^∞. In this paper, we approximate them by Huber function (Huber 1964) which is Lipschitz-C^1 only. Huber function has confirmed its best theoretical approximation with an overall better performance in terms of both the quality of the estimated (Kalmoun 2018) and the speed of convergence. It is a piecewise function defined in (7) with introducing some small threshold $\epsilon > 0$, and it is easy to find that it is non-quadratic but convex. Huber function consists of a quadratic function for arguments below a threshold ϵ (for smoothing small-scale noise) and of a linear function above the threshold ϵ (for preserving discontinuities). So the unconstrained functional with Huberization regularizing terms is as follows

$$J_{\alpha,\epsilon}(f) = \frac{1}{2} \|PF_t f - P\hat{g}^\sigma\|_{l^2}^2 + \alpha_1 \|\Psi \circ f\|_{l^1,\epsilon} + \alpha_2 |f|_{TV,\epsilon}, \tag{12}$$

where

$$\|\Psi \circ f\|_{l^1,\epsilon} = \sum_{m,n=1}^{N} \phi_\epsilon \left((\Psi \circ f)_{l(m,n)} \right) \tag{13}$$

$$|f|_{TV,\epsilon} = \sum_{m,n=1}^{N} \phi_\epsilon \left(|\nabla_{m,n} f| \right), \tag{14}$$

where $l(m,n) = (n-1)N + m$, and operator $\nabla_{m,n} f = \left(\nabla_{m,n}^{x_1} f, \nabla_{m,n}^{x_2} f \right)$, $|\nabla_{m,n} f| = \sqrt{(\nabla_{m,n}^{x_1} f)^2 + (\nabla_{m,n}^{x_2} f))^2}$ with two components defined as

$$\nabla^{x_1}_{m,n} f = \begin{cases} f_{m+1,n} - f_{m,n}, & \text{if } m < N, \\ f_{1,n} - f_{m,n}, & \text{if } m = N, \end{cases} \quad \nabla^{x_2}_{m,n} f = \begin{cases} f_{m,n+1} - f_{m,n}, & \text{if } n < N, \\ f_{m,1} - f_{m,n}, & \text{if } n = N \end{cases}$$

for $m, n = 1, \cdots, N$ due to the periodic boundary condition for f. Hence, our image restoration problem is finally reformulated as the following unconstraint problem

$$\begin{cases} f^* := \arg\min_f J_{\alpha,\epsilon}(f), \\ J_{\alpha,\epsilon}(f) = \frac{1}{2}\|PF_t f - P\hat{g}^\sigma\|^2_{l^2} + \alpha_1\|\Psi \circ f\|_{l^1,\epsilon} + \alpha_2|f|_{TV,\epsilon}. \end{cases} \tag{15}$$

The image restoration using incomplete frequency data based on TV-L^2, l^1-L^2 or TV-L^1 models has been studied for a long time, mainly focused on the efficient iteration algorithms for yielding the local minimizer of the cost functional (Hintermüller and Rincon-Camacho 2010; Zhu et al. 2014). The conjugate gradient method (CGM) (Lustig et al. 2007), gradient project method (Figueiredo et al. 2007), fixed-point continuation method (Hale et al. 2011), split Bregman method (Goldstein and Osher 2009), fast alternating minimization method (Zhu and Chern 2011), the operator-splitting algorithm (OS) (Ma et al. 2008), and fast iterative shrinkage-thresholding algorithm (FISTA) (Beck and Teboulle 2009) are all the efficient approaches to image restoration. Although there already exist several better algorithms recently for finding the local minimizer of the cost functional such as half-quadratic approximation (Yin et al. 2015) and alternating direction method (ADM) (Yang et al. 2010), the direct iteration schemes for multiregularizating model are required further studied to decrease the memory space and computational costs. In the following numerical implementation, we compared the proposed algorithm with ADM for TVl^1-L^2 model in order to demonstrate ours is practical feasible and promising. The ADM is an algorithm based on the classic augmented Lagrangian method (ALM). The convergence of the ADM for the step length was established in the context of variational inequality (Yang et al. 2010).

According to the Bregman iteration method which is established on the theory of Bregman distance (Brègman 1967), to solve the optimization problem for yielding $f^{(k+1)}$ which is implemented by solving its Euler–Lagrange equation (Sun and Yuan 2006), i.e., the iteration framework of Bregman iteration is

$$\begin{cases} \hat{g}^{(k+1)} \leftarrow \hat{g}^\sigma + \left(P\hat{g}^{(k)} - PF_t f^{(k)}\right), \\ f^{(k+1)} \leftarrow \arg\min_f\{\alpha_1\|\Psi \circ f\|_{l^1,\epsilon} + \alpha_2|f|_{TV,\epsilon} + \frac{1}{2}\|PF_t f - P\hat{g}^{(k+1)}\|^2_{l^2}\}. \end{cases} \tag{16}$$

In order to solve the Euler–Lagrange equation of (12), we need the derivatives of data-matching term and each penalty term with respect to the vector $f \in \mathbb{R}^{N^2\times 1}$.

The derivative of data-fitting term in (12) is easy, since the first term is a quadratic function of f. Let us compute the Huberization penalty terms. Firstly, we give the gradient of Huber function, i.e.,

$$\phi'_\epsilon(s) = \begin{cases} \frac{s}{\epsilon}, & |s| \le \epsilon, \\ \frac{s}{|s|}, & |s| > \epsilon. \end{cases} \tag{17}$$

Based on (17), the derivative of sparsity penalty term is as follows

$$
\begin{aligned}
\frac{\partial}{\partial f_{j'(m',n')}}\|\Psi\circ f\|_{l^1,\epsilon} &= \sum_{m,n=1}^{N}\phi'_\epsilon((\Psi\circ f)_{j(m,n)})\cdot\frac{\partial}{\partial f_{j'(m',n')}}(\Psi\circ f)_{j(m,n)} \\
&= \sum_{j=1}^{N^2} a_{j(m,n)}[f]\left(\Psi\circ\frac{\partial}{\partial f_{m',n'}}f\right)_{j(m,n)} \\
&= \sum_{j=1}^{N^2} a_{j(m,n)}[f]\Psi^T(\Psi\circ f)_{j(m,n)}\frac{\partial}{\partial f_{m',n'}}f \\
&= a_{j'(m',n')}[f]f_{j'(m',n')},
\end{aligned}
\tag{18}
$$

To this end, we notice that for any fixed $l\in\{1,2,\cdots,N^2\}$, we have

$$
\frac{\partial}{\partial f_l}\|\Psi\circ f\|_{l^1,\epsilon} = a_l[f]f_l. \tag{19}
$$

Now, consider the TV penalty term. By the representations in our recent work (Liu and Liu 2019), the derivative is in fact for the variable vector f_j is as follows

$$
\begin{aligned}
\frac{\partial}{\partial f_{j'(m',n')}}|f|_{TV,\epsilon} &= \sum_{m,n=1}^{N}\phi'_\epsilon(|\nabla_{m,n}f|)\cdot\frac{\partial}{\partial f_{j'(m',n')}}|\nabla_{m,n}f| \\
&= \sum_{j=1}^{N^2} b_{j(m,n)}[f]\cdot\frac{\nabla_{m,n}f}{|\nabla_{m,n}f|}\cdot\frac{\partial}{\partial f_{j'(m',n')}}\begin{pmatrix}\nabla^{x_1}_{m,n}f\\ \nabla^{x_2}_{m,n}f\end{pmatrix} \\
&= \sum_{j=1}^{N^2}\frac{b_{j(m,n)}[f]}{|\nabla_{m,n}f|}\cdot\nabla_{m,n}f\cdot\begin{pmatrix}(D_r)_{j(m,n),j'(m',n')}\\ (D_c)_{j(m,n),j'(m',n')}\end{pmatrix} \\
&= \sum_{j=1}^{N^2}\left((D_r)_{j,j'},(D_c)_{j,j'}\right)\cdot(\nabla_{m,n}f)^T\cdot\frac{b_j[f]}{|\nabla_{m,n}f|} \\
&= \sum_{j=1}^{N^2}((D_r)_{j,j'},(D_c)_{j,j'})\frac{b_j[f]}{|\nabla_{m,n}f|}\sum_{k=1}^{N^2}\begin{pmatrix}(D_r)_{j,k}\\ (D_c)_{j,k}\end{pmatrix}f_k.
\end{aligned}
\tag{20}
$$

where $D_r := I\otimes D$, $D_c := D\otimes I$ with identity matrix I and circulant matrix $D :=$ **circulant**$(-1,0,0,\cdots,0,1)\in\mathbb{R}^{N\times N}$. Fixed the index as l, the above derivative is rewritten as follows

$$\frac{\partial}{\partial f_l}|f|_{TV,\epsilon} = \sum_{j,k=1}^{N^2}((D_r)_{j,l},(D_c)_{j,l})\cdot((D_r)_{j,k},(D_c)_{j,k})^T \cdot\frac{b_j[f]}{|\nabla_{\left(mod[j,N],int\left[\frac{j-1}{N}\right]+1\right)}f|}f_k$$
$$= \left(\frac{(D_r)_{1,l}}{d_1[f]},\frac{(D_r)_{2,l}}{d_2[f]},\cdots,\frac{(D_r)_{N^2,l}}{d_{N^2}[f]}\right)D_r f + \left(\frac{(D_c)_{1,l}}{d_1[f]},\frac{(D_c)_{2,l}}{d_2[f]},\cdots,\frac{(D_c)_{N^2,l}}{d_{N^2}[f]}\right)D_c f. \tag{21}$$

Combining (19) and (21), the first-order condition for (12) is as follows

$$\begin{cases}\nabla_f \frac{1}{2}\|PF_t f - P\hat{g}^{\sigma}\|_{l^2}^2 = F_t^* P^*(PF_t f - P\hat{g}^{\sigma}),\\ \nabla_f\|\Psi\circ f\|_{l^1,\epsilon} := A[f]f,\\ \nabla_f|f|_{TV,\epsilon} := B[f]f,\end{cases} \tag{22}$$

where

$$\begin{cases}A[f] := \mathrm{diag}(a_{j(m,n)}[f]),\\ B[f] := D_r^T\Lambda[f]D_r + D_c^T\Lambda[f]D_c := B_1[f]+B_2[f]\end{cases}$$

with

$$a_{j(m,n)}[f] := \begin{cases}\frac{(\Psi\circ f)_{j(m,n)}}{\epsilon}, & |(\Psi\circ f)_{j(m,n)}|\le\epsilon,\\ \frac{(\Psi\circ f)_{j(m,n)}}{|(\Psi\circ f)_{j(m,n)}|}, & |(\Psi\circ f)_{j(m,n)}|>\epsilon,\end{cases}$$
$$b_{j(m,n)}[f] := \begin{cases}\frac{|\nabla_{m,n}f|}{\epsilon}, & |\nabla_{m,n}f|\le\epsilon,\\ 1, & |\nabla_{m,n}f|>\epsilon,\end{cases}$$
$$d_{j(m,n)}[f] := \frac{\left|\nabla_{\left(mod[j,N],int\left[\frac{j-1}{N}\right]+1\right)}f\right|}{b_{j(m,n)}[f]} = \frac{|\nabla_{m,n}f|}{b_j[f]}$$
$$\Lambda[f] := \mathrm{diag}\left(\frac{1}{d_1[f]},\frac{1}{d_2[f]},\cdots,\frac{1}{d_{N^2}[f]}\right)$$

for $j=1,\cdots,N^2$ which defined as $j := j(m,n) = (n-1)N+m$ for $m,n = 1,\cdots,N$. Notice, the specified $B_1[f]$, $B_2[f]$ are $N^2\times N^2$ symmetric positive definite matrices.

With (22) and input frequency data $\hat{g}^{\sigma} := \hat{g}^{(k+1)}$, we know that $f^{(k+1)}$ could be solved by the nonlinear Euler–Lagrange equation of the model (12),

$$F_t^*P^*(PF_t f - P\hat{g}^{(k+1)}) + \alpha_1 A[f]f + \alpha_2 B[f]f = 0, \tag{23}$$

that is

$$F_t^*P^*PF_t f + \alpha_1 A[f]f + \alpha_2 B[f]f = F_t^*P^*P\hat{g}^{(k+1)}. \tag{24}$$

To solve numerically (24) is not a easy task because of the presence of the nonlinear term $A[f]$ and diffusivity coefficient $\Lambda[f]$ in $B[f]$. Using the fixed-point iteration scheme proposed by Vogel and Oman (Vogel and Oman 1996) could change (24) to a linear equation:

$$F_t^* P^* P F_t f^{(k+1)} + \alpha_1 A[f^{(k)}] f^{(k+1)} + \alpha_2 B[f^{(k)}] f^{(k+1)} = F_t^* P^* P \hat{g}^{(k+1)}. \quad (25)$$

Let $\mathbf{C} := \alpha_1 A[f^{(k)}] + \alpha_2 B[f^{(k)}] \in \mathbb{C}^{N^2 \times N^2}$, then (25) could be rewritten as

$$\mathbf{C} f^{(k+1)} + F_t^* P^* P F_t f^{(k+1)} = F_t^* P^* P \hat{g}^{(k+1)}. \quad (26)$$

Multiply both sides of (26) by the two-dimensional discrete Fourier matrix F_t and obtain

$$F_t \mathbf{C} f^{(k+1)} + P^* P F_t f^{(k+1)} = P^* P \hat{g}^{(k+1)}. \quad (27)$$

According to the fast algorithm in Liu and Zhu (2014), $A[f^{(k)}]$ is diagonal and $B[f^{(k)}]$ is block circulate with circulate blocks (BCCB) (Vogel 2002). Denoted

$$\begin{aligned} \mathbf{C} &= \mathbf{circulant}(C_1, C_2, \cdots, C_N) \\ &:= \mathbf{bccb} \begin{pmatrix} c_1^1 & c_1^2 & \cdots & c_1^N \\ c_2^1 & c_2^2 & \cdots & c_2^N \\ \vdots & \vdots & \ddots & \vdots \\ c_N^1 & c_N^2 & \cdots & c_N^N \end{pmatrix} = \mathbf{bccb}(\dot{C}). \end{aligned}$$

where $C_j = \mathbf{circulant}(c_1^j, c_2^j, \cdots, c_N^j)$. Then using the property in Vogel (2002), i.e.,

$$F_t \mathbf{C} f^{(k+1)} = \mathbf{D} F_t f^{(k+1)},$$

where $\mathbf{D} \in \mathbb{C}^{N^2 \times N^2}$ is a diagonal matrix with components $\mathbf{D}_{j,j} = (F_t \dot{C})_j$, $j = 1, \cdots, N^2$. Therefore, (27) becomes

$$(\mathbf{D} + P^* P) F_t f^{(k+1)} = P^* P \hat{g}^{(k+1)}. \quad (28)$$

Since $P^* P$ is a diagonal matrix because of the definition, $F_t f^{(k+1)}$ is easily obtained by (28), so is $f^{(k+1)}$.

Due to the input frequency data that have the degeneration with relative random noise, we consider to use the spatial filter as the preprocessing step before iteration. Based on the fast algorithm in Liu and Zhu (2014) to solve minimizer of $J_{\alpha,\epsilon}(f)$, we could obtain the minimizer of functional $J_{\alpha,\epsilon}(f)$ with preprocessing on data

$$\tilde{g}(\boldsymbol{\omega}) = \mathcal{F} \circ \tilde{g}(\boldsymbol{x}) := \mathcal{F} \circ [G_\tau * g(\boldsymbol{x})], \quad (29)$$

where τ is the standard deviation of the spatial filter with the filtering window size T. That means we use the discrete form $\tilde{g}^{\sigma}_{m',n'}$ instead of $\hat{g}^{\sigma}_{m',n'}$. Let $\tilde{f}^*$ be the minimizer of functional $J_{\alpha,\epsilon}(f)$ with preprocessing. Then the fast algorithm process for the minimizing problem (15) with preprocessing can be described in algorithm. With the preprocessing step, it could avoid the aliasing frequency and make the reconstructed image clearly.

In the next section, we describe numerical experiments that demonstrate that the proposed algorithm is very efficient for MRI image restoration with high-level relative random noise.

3 Numerical Experiments

In this section, the performance of algorithm in solving model (15) for TVl^1-L^2 MRI image restoration is evaluated. The proposed method is compared with the fast iteration algorithm in Liu and Zhu (2014).

The signal-to-noise ratio (SNR) and relative error (ReErr) are used to measure the quality of the reconstructed images. The definitions of SNR and ReErr are given as follows:

$$\mathrm{SNR} = 20\lg\left(\frac{\|f\|^2_{l^2}}{\|f - f^{(k)}\|^2_{l^2}}\right), \quad \mathrm{ReErr} = \frac{\|f^{(k)} - f\|^2_{l^2}}{\|f\|^2_{l^2}}, \tag{30}$$

Algorithm A fast scheme for Huberization image restoration model

Input: frequency noisy data $\{\hat{g}^{\sigma}_{m',n'} : m', n' = 1, \cdots, N\}$, sampling matrix $P \in \mathbb{R}^{N^2 \times N^2}$, parameters $K_0, \alpha, \epsilon, \tau$ and filtering window size T.
Preprocessing: $\tilde{g}^{\sigma}_{m',n'} = \mathcal{F} \circ [G_\tau * \mathcal{F}^{-1} \circ \hat{g}^{\sigma}_{m',n'}]$.
Do the iteration from $k = 0, 1, \ldots$ with $\hat{g}^{(0)} = \mathbf{0}$, $f^{(0)} = \mathbf{0} \in \mathbb{R}^{N^2 \times 1}$.
While $k < K_0$
{ Compute:
    $PFf^{(k)}$,
    $\hat{g}^{(k+1)} \leftarrow \tilde{g}^{\sigma} + (P\hat{g}^{(k)} - PF_t f^{(k)})$,
    $P^* P\hat{g}^{(k+1)}$, $P^* P$, $\mathbf{D}$,
    $f^{(k+1)}$ by (28). }
End while
$\tilde{f}^* := f^{(K_0)}$.
End

where $f^{(k)}$ and f are the reconstructed and original images, respectively. The CPU time is used to evaluate the speed of MRI restoration. All numerical implementations are performed in MATLAB R2017b on a laptop with 1.6GHz Intel Core i5 processor and 8 GB of memory.

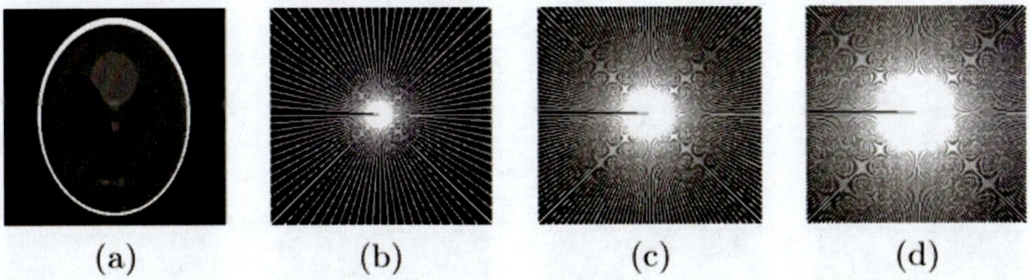

Fig. 1 **a** Original image; **b** sampling mask: 22 × 2 lines; **c** 22 × 4 lines; **d** 22 × 6 lines

Firstly, we yield the full noisy data $g^{\sigma}_{m,n}$ from the exact image $f_{m,n}$ by the scheme

$$g^{\sigma}_{m,n} = f_{m,n} + \sigma \times rand(m,n) \times f_{m,n},$$

where $m, n = 1, \cdots, N$, and $rand(m,n)$ are the random numbers in $[-1, 1]$. Then the input frequency data are $\hat{g}^{\sigma}_{m',n'} = \mathcal{F} \circ g^{\sigma}_{m,n}$. It is easy to see that this is a relative random noise in spatial domain.

The choice strategy of double regularizing parameters is proposed in the recent paper (Liu and Liu 2019). In this work, it will not be introduced in detail. Let $\alpha_1 = 0.001$, $\alpha_2 = 0.001$, and $\epsilon = 0.01$ are in reconstruction tests. In the following numerical tests, the radial sampling method with K sampling lines is used as sampling matrix P. The sampling ratio is K/N^2. The stopping criterion is the maximum iteration number $K_0 = 100$. The Gaussian filter is be chosen as the filter in preprocessing.

Example 1 [Reconstruction for phantom image with different parameters]
We use a 256 × 256 smooth piecewise phantom image shown in Fig. 1a as the initial image f. We test the fast algorithm with radial sampling method with 22 × 2, 22 × 4, 22 × 6 lines shown in Fig. 1b–d, i.e., the sampling ratio is 18.76%, 34.97%, 49.68%, respectively. The other parameters are the same above, i.e., $\alpha_1 = \alpha_2 = 0.001$, $\epsilon = 0.01$, $K_0 = 100$. The noisy image Fig. 2a is being added relative noise level $\sigma = 0.5$, i.e., added 50% relative random noise on the initial image. The first line in Fig. 2 is the reconstruction test under 22 × 2 sampling lines, while the second and third lines are under 22 × 4 and 22 × 6 sampling lines, respectively. So the restoration image under different sampling without preprocessing is shown in Fig. 2b. And Fig. 2c, d is the images using the proposed algorithm with the standard deviation of the Gaussian filter $\tau = 3$ and filtering window size $T = 3, 5$, respectively.

Denote the parameters of Gaussian filter as a pair array (τ, T). We do the numerical experiments under proposed algorithm with different preprocessing. The test results are shown in Table 1, with the best one in bold. From the these data, even though the CPU time is bigger (that because of the preprocessing step), the SNR in proposed algorithm is much bigger than the compared method in Liu and Zhu (2014), and relative error is smaller than the one without preprocessing. It is clearly that the

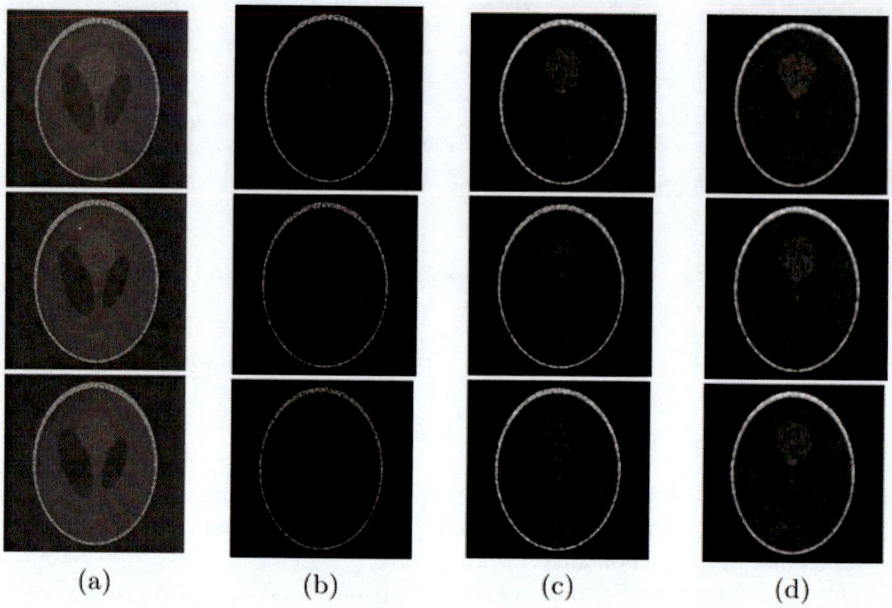

Fig. 2 Reconstructions from 50% noisy image under 22 × 2, 22 × 4, 22 × 6 sampling lines. From left to right: **a** noisy image; **b** restoration image without preprocessing; **c** with preprocessing $\tau = 3$ and filter window $T = 3$; **d** with preprocessing $\tau = 3$ and filter window $T = 5$

restoration images in Fig. 2c, d have clearer edges and areas than Fig. 2b. In addition, the more sampling lines are, the clearer the restoration image is.

Now, let noise level be $\sigma = 0.7, 0.9$, i.e., added 70%, 90% relative random noise on the same brain image, respectively (shown in the second, third line in Fig. 3a). The sampling method is the radial sampling with 22 × 6 sampling lines, and other parameters are the same as above. The restoration images in compared and proposed method are shown in the second and third lines in Fig. 3b–d. The partial results data are shown in Table 2, with the best one in bold.

From the results, it is clear that the compared method could not reconstruct the multiplicative noisy data even though the sampling ratio is big enough, while the proposed method could do better, especially to the edge-preservations and areas recognition. On the other hand, the optimal filter parameters are not the same. Hence, the choice of standard deviation and filter window size could be one of the future works.

Example 2 [Comparison for MRI image restoration with relative random noise]
Now different size of magnetic resonance imaging (MRI) images are shown in the first line of Fig. 4. There are 256 × 256 MR brain image, 256 × 256 brain section bitmap, 512 × 512 articular born image, 220 × 220 internal organs image, and 220 × 220 blood vessels image. In real application, the patients could not stay in the nuclear

Table 1 Test data of 50% relative noisy image with 22 × 2, 22 × 4, 22 × 6 sampling in compared method and proposed method with filter parameters (τ, T), with the best one in bold

Paras/22 × 2	**SNR(dB)**	**ReErr**	**CPU time(s)**
without	8.8409	0.3614	3.8322
(3, 3)	**11.0048**	**0.2817**	3.2184
(3, 5)	9.6254	0.3302	3.2426
(5, 3)	10.8662	0.2862	3.2536
(5, 5)	9.5368	0.3335	3.1957
Paras/22 × 4	**SNR(dB)**	**ReErr**	**CPU time(s)**
without	6.9877	0.4473	4.0658
(3, 3)	**11.0416**	**0.2805**	3.2675
(3, 5)	9.5856	0.3317	3.2370
(5, 3)	10.9192	0.2845	3.2513
(5, 5)	9.3566	0.3405	3.2380
Paras/22 × 6	**SNR(dB)**	**ReErr**	**CPU time(s)**
without	6.2300	0.4881	4.8800
(3, 3)	**10.6467**	**0.2935**	3.6304
(3, 5)	9.6451	0.3294	3.7328
(5, 3)	10.6293	0.2941	4.2108
(5, 5)	9.3391	0.3412	4.3454

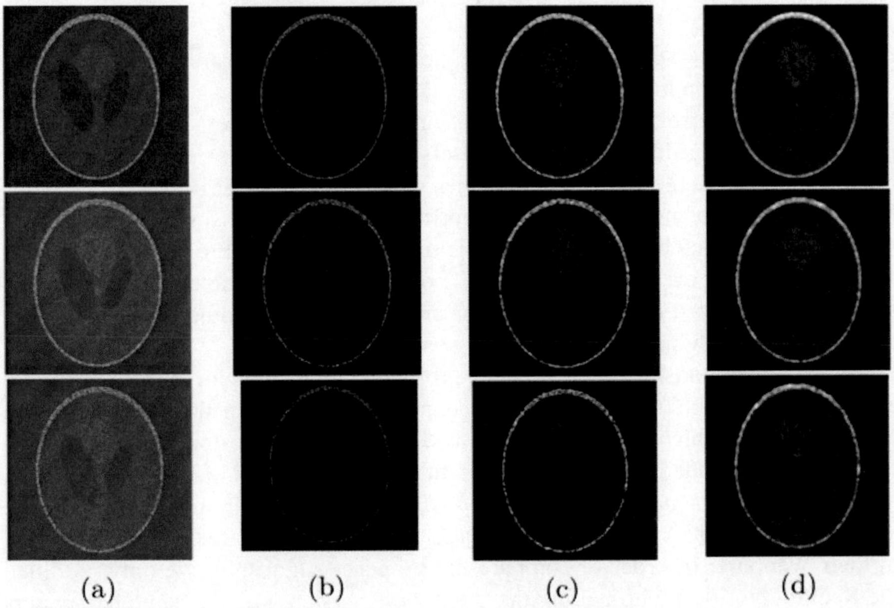

Fig. 3 Reconstructions from 50%, 70%, 90% noisy image under 22 × 6 sampling lines. From left to right: **a** noisy image; **b** restoration image without preprocessing; **c** with preprocessing $\tau = 3$ and filter window $T = 3$; **d** with preprocessing $\tau = 3$ and filter window $T = 5$

Table 2 Test data of 50%, 70%, 90% relative noisy image with 22 × 6 sampling in compared method and proposed method with filter parameters (τ, T), with the best one in bold

Paras/50% noise	**SNR(dB)**	**ReErr**	**CPU time(s)**
without	6.2300	0.4881	4.8800
(3, 3)	**10.6467**	**0.2935**	3.6304
(3, 5)	9.6451	0.3294	3.7328
(5, 3)	10.6293	0.2941	4.2108
(5, 5)	9.3391	0.3412	4.3454
Paras/70% noise	**SNR(dB)**	**ReErr**	**CPU time(s)**
without	4.2459	0.6133	4.2390
(3, 3)	9.6967	0.3275	4.1380
(3, 5)	9.2002	0.3467	4.1169
(5, 3)	**9.7378**	**0.3259**	4.3043
(5, 5)	9.0231	0.3539	4.1071
Paras/90% noise	**SNR(dB)**	**ReErr**	**CPU time(s)**
without	1.9137	0.8023	4.0034
(3, 3)	8.4034	0.3800	4.1857
(3, 5)	**8.7580**	**0.3648**	4.1252
(5, 3)	8.3189	0.3838	3.9400
(5, 5)	8.5200	0.3749	4.0363

medicine instrumentation for a long time. So the sampling data are partial noisy data which collected in a little time.

We added 90% relative random noise on it to simulate the MRI and the other parameters are the same, i.e., the parameters $\alpha_1 = \alpha_2 = 0.001$, $\epsilon = 0.01$, and using radial sampling method with 22 × 6 lines. The second line in Fig. 4 is the noisy MRI images. By using the proposed algorithm with preprocessing step, the reconstructed images are obtained shown in the third line of Fig. 4. From the figures, it is easy to know that even though the details of each image in the third line of Fig. 4 are not reconstructed clearer, the patterns are much efficient to be recognized from multiplicative noisy images.

On the other hands, we compared the proposed algorithm with the fast algorithm called reconstruction from partial Fourier data (RecPF) based on alternating direction method (ADM), which is the main efficient scheme for image restoration today (Yang et al. 2010). We added 90% relative random noise on the MRI brain images Fig. 5a, respectively, i.e., the noisy images are shown in Fig. 5b. The parameters are the same as the one chosen in Yang et al. (2010), and the algorithm for RecPF is downloaded in their Web site. In order to compare the proposed algorithm, we only modified the RecPF algorithm with adding the preprocessing step, i.e., the same parameters as above ($\tau = 3$ and filter window $T = 5$). Figure 5c, d are the reconstructions by RecPF and ours.

The SNR for MRI brain images by RecPF is 11.1761 dB and 11.7596 dB, while the SNR by our proposed scheme is 12.5751 dB and 12.4399 dB, respectively. Even

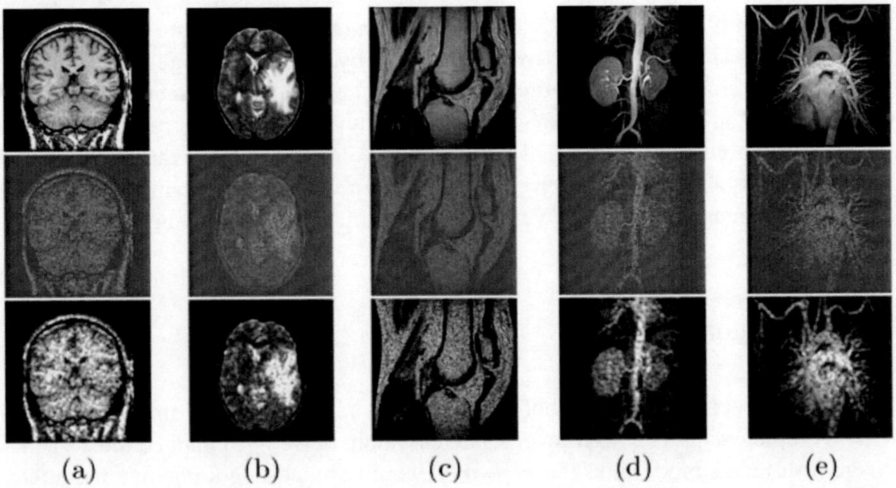

Fig. 4 Reconstructions from 90% noisy images under 22 × 6 sampling lines. From up to bottom: initial MRI images, noisy images, reconstructed images with preprocessing $\tau = 3$, and filter window $T = 5$

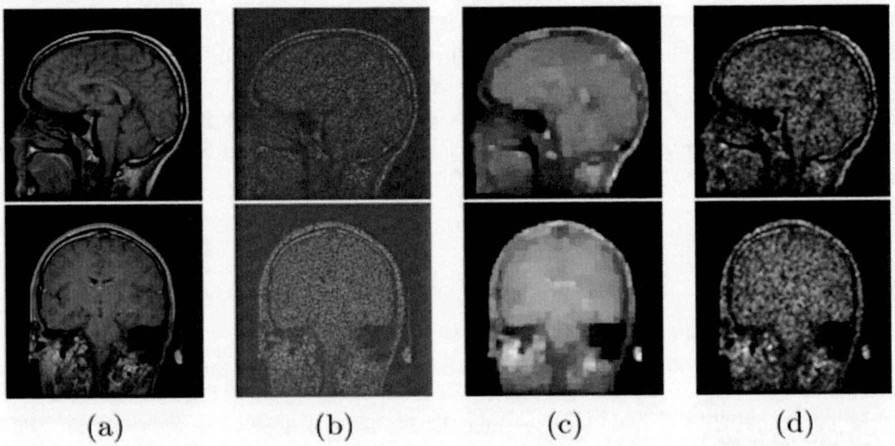

Fig. 5 Reconstructions from 90% noisy image under 22 × 6 sampling lines. From left to right: **a** initial MRI brain images; **b** 90% noisy images; **c** restored image by RecPF; **d** restored image by the proposed scheme

though the sampling chosen 22×6 lines to guarantee that there are more efficient frequency data are used, the reconstructed image by RecPF has large error occurred in both the interior part and the edge of the image, i.e., the reconstructed image could not be seen as an effective and feasible result. Anyway, the reconstructed image by the proposed algorithm with small error and noise point could be the effective and efficient one. It shows that the proposed algorithm with preprocessing step is more efficiently to image restoration even though the noise pollution is huge.

4 Conclusion

We added a preprocessing step before iteration to restore the MRI image from limited incomplete data with high-level relative random noise. We changed the TVl^1-L^2 image restoration model into the perturbed version by adding a positive parameter ϵ. Bregman iteration was used to solve the modified model with double Huberization penalty terms. The proposed method was compared with fast algorithm using two FFTs and one preprocessing step. A smooth piecewise image and some MRI images with high-level relative random noise are employed to test in the numerical experiments. The results demonstrate that the proposed method is very efficient to reconstruct image with this kind of multiplicative noise by using limited noisy data.

Notes and Comments: This work is supported by the National Natural Science Foundation of China (Grant No. 11531005, 11871149), the Natural Science Foundation of Jiangsu Province (Grant No. BK20210380).

References

Aubert G, Vese L (1997) A variational method in image recovery. SIAM J Numer Anal 34(5):1948–1979

Aubert G, Kornprobst P (2006) Mathematical problems in image processing, 2nd edn. Appl Math Sci 147 (Springer, New York)

Beck A, Teboulle M (2009) A fast iterative shrinkage-thresholding algorithm for linear inverse problems. SIAM J Imaging Sci 2(1):183–202

Bredies K, Kunisch K, Pock T (2010) Total generalized variation. SIAM J Imag Sci 3(3):492–526

Brègman LM (1967) The relaxation method of finding the common point of convex sets and its application to the solution of problems in convex programming. USSR Comput Math Math Phys 7(3):200–217

Candès EJ, Romberg J, Tao T (2006) Robust uncertainty principles: exact signal reconstruction from highly incomplete frequency information. IEEE Trans Inf Theor 52(2):489–509

Chan TF, Golub GH, Mulet P (1999) A nonlinear primal-dual method for total variation-based image restoration. SIAM J Sci Comput 20(6):1964–1977

Chen SS, Donoho DL, Saunders MA (2001) Atomic decomposition by basis pursuit. Siam Rev 43(1):129–159

Clason C, Jin BT, Kunisch K (2010) A duality-based splitting method for l^1-TV image restoration with automatic regularization parameter choice. SIAM J Sci Comput 32(3):1484–1505

Daubechies I, Defrise M, De Mol C (2004) An iterative thresholding algorithm for linear inverse problems with a sparsity constraint. Commun Pure Appl Math 57(11):1413–1457

Donoho DL (2006) Compressed sensing. IEEE Trans Inf Theor 52(4):1289–1306

Fan YR, Huang TZ, Liu J et al (2016) Compressive sensing via nonlocal smoothed rank function. PLoS One 11(9):e0162041

Figueiredo MAT, Nowak RD, Wright SJ (2007) Gradient projection for sparse reconstruction: application to compressed sensing and other inverse problems. IEEE J Sel Top Sig Process 1(4):586–597

Goldstein T, Osher S (2009) The split Bregman method for L1-regularized problems. SIAM J Imaging Sci 2(2):323–343

Guo XX, Li F, Ng MK (2009) A fast l^1-TV algorithm for image restoration. SIAM J Sci Comput 31(3):2322–2341

Hale ET, Yin WT, Zhang Y (2007) A fixed-point continuation for l1-regularization with application to compressed sensing. Technical Report TR07-07, Rice University CAAM (2007)

Hintermüller M, Stadler G (2006) An infeasible primal-dual algorithm for total bounded variation-based inf-convolution-type image restoration. SIAM J Sci Comput 28(1):1–23

Hintermüller M, Wu T (2013) Nonconvex TV^q-models in image restoration: analysis and a trust-region regularization-based superlinearly convergent solver. SIAM J Imag Sci 6(3):1385–1415

Hintermüller M, Rincon-Camacho MM (2010) Expected absolute value estimators for a spatially adapted regularization parameter choice rule in L1-TV-based image restoration. Inverse Prob 26(8):085005

Hintermüller M, Stadler G (2003) A semi-smooth newton method for constrained linear-quadratic control problems. Zamm J Appl Math Mech 83(4):219–237

Huber PJ (1964) Robust estimation of a location parameter. Ann Math Stat 35(1):73–101

Kalmoun EM (2018) An investigation of smooth TV-like regularization in the context of the optical flow problem. J Imag 4(2):31

Knoll F, Bredies K, Pock T, Stollberger R (2011) Second order total generalized variation (TGV) for MRI. Magn Reson Med 65(2):480–491

Kunisch K (2004) Total bounded variation regularization as a bilaterally constrained optimization problem. SIAM J Appl Math 64(4):1311–1333

Liu XM, Liu JJ (2019) On image restoration from random sampling noisy frequency data with regularization. Inverse Prob Sci Engi 27(12):1765–1789

Liu XM, Zhu YG (2014) A fast method for TV-L1-MRI image reconstruction in compressive sensing. J Comput Inf Syst 10(2):691–699

Lustig M, Donoho D, Pauly JM (2007) Sparse MRI: the application of compressed sensing for rapid MR imaging. Magn Reson Med 58(6):1182–1195

Ma TH, Lou YF, Huang TZ (2017) Truncated l^{1-2} models for sparse recovery and rank minimization. SIAM J Imag Sci 10(3):1346–1380

Madsen K, Nielsen HB (1993) A finite smoothing algorithm for linear L1 estimation. SIAM J Optim 3(2):223–235

Mallat S (2009) A wavelet tour of signal processing, 3rd edn. Elsevier/Academic Press, Amsterdam

Ma S, Yin WT, Zhang Y, Chakraborty A (2008) An efficient algorithm for compressed MR imaging using total variation and wavelets. In: IEEE conference on computer vision and pattern recognition, pp 1–8 (2008)

Rudin LI, Osher S, Fatemi E (1992) Nonlinear total variation based noise removal algorithms. Phys D Nonlinear Phenom 60(1):259–268

Sun WY, Yuan YX (2006) Optimization theory and methods, Springer optimization and its applications, vol 1. Springer, New York

Tibshirani R (1996) Regression shrinkage and selection via the lasso. J R Stat Soc 58(1):267–288

Vogel CR (2002) Computational methods for inverse problems. Frontiers Appl Math 23 (Society for Industrial and Applied Mathematics (SIAM), Philadelphia, PA)

Vogel CR, Oman ME (1996) Iterative methods for total variation denoising. SIAM J Sci Comput 17(1):227–238

Yang JF, Zhang Y, Yin WT (2010) A fast alternating direction method for TVL1-L2 signal reconstruction from partial Fourier data. IEEE J Sel Top Signal Process 4(2):288–297

Yin WT, Osher S, Goldfarb D et al (2008) Bregman iterative algorithms for l1-minimization with applications to compressed sensing. SIAM J Imag Sci 1(1):143–168

Yin PH, Lou YF, He Q et al (2015) Minimization of l^{1-2} for compressed sensing. SIAM J Sci Comput 37(1):A536–A563

Zhu YG, Liu XM (2015) A fast method for L1–L2 modeling for MR image compressive sensing. J Inverse Ill-Posed Prob 23(3):211–218

Zhu YG, Shi YY (2013) A fast method for reconstruction of total-variation MR images with a periodic boundary condition. IEEE Sig Process Lett 20(4):291–294

Zhu YG, Shi YY, Zhang B et al (2014) Weighted-average alternating minimization method for magnetic resonance image reconstruction based on compressive sensing. Inverse Prob Imag 8(3):925–937

Zhu Y, Chern I (2011) Fast alternating minimization method for compressive sensing MRI under wavelet sparsity and TV sparsity. In: Proceedings of 2011 sixth international conference on image and graphics, pp 356–361 (2011)

A Brief Review of Some Swarming Models Using Stochastic Differential Equations

Linh Thi Hoai Nguyen, Ton Viet Ta, and Atsushi Yagi

1 Introduction

This study is motivated by the spectacular swarm behavior of animals in the natural world. Our objective is twofold. The first one is to get understanding about the dynamics of swarming in nature. The second one is using information acquired from the study of biological swarming to design artificial/information systems, such as reactive robotic systems, cellular networks, collision-avoiding systems for automobiles.

Animal swarming is known as one of typical self-organization observed phenomenon that is coherently performed by integration of interactions among constituent individuals in biological systems: fish schooling, bird flocking, and mammal herding. It has attracted the interests of researchers from diverse fields including biology, physics, mathematics, computer engineering.

Let us recall some studies in the literature. Empirical study on animal swarming has been done in (Aoki 1982; Breder 1951, 1959; Cullen et al. 1965; Huth and Wissel 1992; Keenleyside 1995; Parrish and Viscido 2005). Based on experimental results, Camazine et al. presented the three basic behavioral rules among individuals in a group (Camazine et al. 2001). Their insight is that these local rules can altogether create the coherent behavior of animal swarms. As for the theoretical approach, we want to quote (D'Orsogna et al. 2006; Oboshi et al. 2002; Olfati-saber 2006; Vicsek et al. 1995). Vicsek et al. (1995) introduced a simple difference model, assuming that each particle is driven with a constant absolute velocity and the average direction of motion of the particles in its neighborhood together with some random perturba-

L. T. H. Nguyen (✉)
Institute of Mathematics for Industry, Kyushu University, 744 Mootoka, Fukuoka 819-0395, Japan
e-mail: linh@imi.kyushu-u.ac.jp

T. V. Ta
Faculty of Agriculture, Center for Promotion of International Education and Research, Kyushu University, 744 Mootoka, Fukuoka 819-0395, Japan

A. Yagi
Department of Applied Physics, Graduate School of Engineering, Osaka University, Suita, Osaka 565-0871, Japan

J. Cheng et al. (eds.), *Proceedings of the Forum "Math-for-Industry" 2018*,
Mathematics for Industry 35, https://doi.org/10.1007/978-981-16-5576-0_9

tion. Oboshi et al. (2002) presented another difference model in which an individual selects one basic behavioral pattern from four based on the distance between it and its nearest neighbor. Meanwhile, Olfati-Saber (2006) and D'Orsogna et al. (2006) independently constructed deterministic differential models using a generalized Morse and attractive/repulsive potential functions, respectively.

In our study, we use the stochastic differential equation (SDE) model approach. Such a model can describe the animal's behavior precisely. Moreover, an SDE model is tractable for making numerical simulations. In this paper, we will use the Euler scheme for stochastic differential equations which has been introduced by Kloeden and Platen (2005).

Let us now review our work in this direction. An SDE model describing the process of schooling is presented in Uchitane et al. (2012), where we use the three behavioral rules proposed in Camazine et al. (2001). We then utilized the model for developing quantitative arguments on fish schooling in Nguyen et al. (Nguyen et al. 2014). The paper (Nguyen et al. 2016) is devoted to studying obstacle-avoiding patterns and cohesiveness of fish school. A foraging model is recently studied in Ta and Nguyen (2008). The result of some of these papers is preliminarily presented in the first author's Doctoral Dissertation (Nguyen 2014).

The group cohesiveness, which is a quantity measuring the internal strength of animal swarm or group, has already been introduced since the 1930s. Moreno and Jennings (1937) defined cohesiveness as the forces holding the individuals within the group to which they belong. Not until 1950 was a systematic theory of group cohesiveness constructed by Festinger et al. (1950). Their definition of cohesiveness is "We shall call the total field of forces which act on members to remain in the group the 'cohesiveness' of the group." Gross and Martin (1952) claimed that this definition is inadequate, and they proposed an alternative definition as the resistance of a group to disruptive forces. Carron (1980) defined cohesiveness to be the adhesive property of a group. The study during long years seems to show that it is not an easy problem to define a concept of the cohesiveness precisely and consistently. It has been conceptualized in various ways, but each was based on intuitive assumptions and interpretations.

Experimental study shows that animal benefits from forming a swarm. The work of Gotmark et al. (1986) shows that the foraging success of gulls (Larus ridibundus) increases with flock size up to at least eight birds. In Berdahl et al. (2013), experiments in a shallow tank with a school of 2^n-golden shiner fish (Notemigonus crysoleucas) ($n = 1, 2, \ldots, 6$) were performed. In the experiments, the fish tracked their preferred, darker regions of a circular patch (darkest at its center and transitioning to the brightest light levels) and moved at a constant speed in the tank. It was shown that when school size increased, so did the time-averaging darkness level at the locations of the fish. In other words, large schools track to a target better than smaller ones. However, this benefit of swarming for foraging does not increase forever as a function of swarm size. Experimental evidence on zebrafish demonstrates that the responsiveness of fish school to food smell decreases as school size exceeds an optimal value (Steele et al. 1991).

The aim of this paper is to give a unified framework for our study on the stochastic differential equations models for describing the swarming behavior in the mentioned above studies. Even though in our previous work, we use the term "fish schooling" instead of "animal swarming," we want to emphasize that our models work well for swarming of animal in general. Therefore, in this paper, we use the term "animal swarming" to refer to the phenomena under consideration.

The outline of the paper is as follows. In Sect. 2, we firstly review the general stochastic differential equation model for describing the movement of individual animals together with their mates. The external force factor in the general model is then made specified to adapt to different environments: the free space, the space with obstacle, and the space with both obstacles and food resource. The particular rules for these environments are presented precisely in Sects. 2.2, 2.3, and 2.4. In Sect. 3, we present our numerical study on the three models. The paper ends with some conclusion remarks.

The highlights of the paper are as follows.

- A scientific definition for swarm cohesiveness which is an internal property characterizing the strength of animal swarm is introduced. A method for numerically measuring swarm cohesiveness is proposed.
- On the basis of avoiding obstacle model, we find that there are four clear avoiding patterns, i.e., Rebound, Pullback, Pass and Reunion, Separation. Furthermore, the emerging patterns changes when one control parameter is tuning while the other parameters are kept constant.
- The relationships between swarm cohesiveness, obstacle-avoiding patterns and model parameters are investigated. This shows that swarm cohesiveness can be measured quantitatively through obstacle-avoiding patterns.
- By means of numerical simulation, we verify the experimental studies that animals benefit from swarming for foraging. More precisely, one single individual could hardly find the food resource, while a group of individuals together forming a swarm has better ability to reach the food resource. The probability of successful foraging increases as the number of individuals in the group increases to some critical size. After that, the success probability starts to decrease.

2 Model Equations

2.1 General Model

We have introduced the SDE model

$$\begin{cases} dx_i(t) &= v_i dt + \sigma_i dw_i(t), \quad i = 1, 2, \ldots, N, \\ dv_i(t) &= \Big[-\alpha \sum_{j=1,\, j \neq i}^{N} \left(\frac{r^p}{\|x_i - x_j\|^p} - \frac{r^q}{\|x_i - x_j\|^q} \right) (x_i - x_j) \\ & \quad -\beta \sum_{j=1,\, j \neq i}^{N} \left(\frac{r^p}{\|x_i - x_j\|^p} + \frac{r^q}{\|x_i - x_j\|^q} \right) (v_i - v_j) \\ & \quad +F_i(t, x_i, v_i)\Big] dt, \quad i = 1, 2, \ldots, N, \end{cases} \tag{1}$$

for describing the process of swarming of N-individual system. In building up such a stochastic differential equation model, we have referred to the individual's behavioral rules introduced by Camazine et al. ((Camazine et al., 2001, Chap. 11)). We have also referred to the idea due to Reynolds (1987). For the details, the reader can consult the paper (Uchitane et al. 2012).

Each individual is regarded as a moving particle in the Euclidean space $\mathbb{R}^d$, where $d = 2$ or 3. The unknown $x_i(t)$ and $v_i(t)$ are stochastic processes with values in $\mathbb{R}^d$ denoting the position and velocity of the ith individual at time t, respectively.

The first equation of (1) is a stochastic one concerning x_i. The noise term $\sigma_i dw_i$ here models a degree of uncertainty in the individuals' behavior that reflects both the imperfect information-gathering ability and the imperfect execution of the individual's actions. In fact, $\{w_i(t),\ t \geqslant 0\}$ ($i = 1, 2, \ldots, N$) are independent d-dimensional Brownian motions defined on a complete probability space with filtration $(\Omega, \mathcal{F}, \{\mathcal{F}_t\}_{t \geqslant 0}, \mathbb{P})$ satisfying the usual conditions.

The second equation is a deterministic one on v_i, where $1 < p < q < \infty$ are fixed exponents, $r > 0$ is a fixed distance, and $\alpha,\ \beta$ are positive coefficients for interaction between individuals and velocity matching, respectively. Each one tries to adjust its position relative to others in the group by repulsive or attractive forces depending on whether the distance to each of its neighbor is less than or greater than the critical distance r. It also attempts to match its velocity to those of the others.

In order to model the attractive and repulsive force, we use a generalization of the Van der Waals force in molecular physics, which is a distance-dependent interaction between atoms and molecules. The exponent p is concerned with the range of attraction among individuals. If p is small, then the attractive force reaches a wide range beyond the critical distance r. In contrast, if p is large, then the attraction is available only in a neighborhood of the disk of radius r.

Finally, the functions $F_i(t, x_i, v_i)$ denote external forces at time t which are given functions defined for (x_i, v_i) with values in $\mathbb{R}^d$.

From the model equation, we have that

$$\sum_{i=1}^{N} dv_i = \sum_{i=1}^{N} F_i(t, x_i, v_i) dt. \tag{2}$$

That is, the sum of increments of all velocities equals to that of all external forces.

In the following subsections, we discuss specialized models in different environments.

2.2 Model in Free Space

For this model, we let the group of animal move in the unbounded, continuous and homogeneous space $\mathbb{R}^d$. We take $F_i(t, x_i, v_i) = -kv_i$ which is usually used to present the resistance in physical particle systems. For example in the case of fish schooling, this force describes the friction force from the water resulting from fish movement.

Furthermore, if we initialize the system from no transport, namely the initial velocities of all individuals in the group are zeros, then from (2),

$$\sum_{i=1}^{N} dv_i = -c\left(\sum_{i=1}^{N} v_i\right) \mathrm{d}t.$$

Consequently, $\sum_{i=1}^{N} dv_i$ decays exponentially as $t \to \infty$ and the system converges to a steady state.

In the real world, the environment surrounding animals often include some components such as obstacles and food resources. In those situations, animals exhibit more complex, parallel movements while avoiding obstacles and finding food resource. We deal with these behavioral rules in the next two subsections.

2.3 Avoiding Obstacle Model

We want to study (1) from the viewpoint of pattern formation. Pattern formation is observed very often in self-organizing systems. What is interesting is, as generally known, that a single mechanism of self-organization can create various patterns by parameter tuning or template regulation. We shall show that the pattern solutions of (1) also have this nature.

As we can observe easily in the natural world that when a group of individuals is tackled by obstacles, the individuals will react quickly to avoid collision with the obstacle. In order to study this behavior using our SDE model, we set a global obstacle in the space where individuals move and then investigate obstacle avoidance patterns performed by the group. In addition to the individual–individual interaction rules, a behavioral rule to avoid collision with the obstacle is included.

Let us consider a spherical obstacle in the $\mathbb{R}^d$ space with central point x_C and radius $\rho > 0$. Therefore, the individuals can move in the domain

$$\Omega = \{x \in \mathbb{R}^d : \|x - x_C\| > \rho\}.$$

The surface of the obstacle is denoted by $S = \{x \in \mathbb{R}^d : \|x - x_C\| = \rho\}$.

The obstacle avoiding rule is that each individual executes an action for avoiding collision with the obstacle according to a reflection law of velocity with distance-depending weights. The rule for avoiding obstacles is included in the external forces.

$$F_i(t, x_i, v_i) = -\gamma \left(\frac{R^P}{\|x_i - y_i\|^P} + \frac{R^Q}{\|x_i - y_i\|^Q} \right) (v_i - u_i),$$

where R is the critical distance to the obstacle and $1 < P < Q < \infty$ are fixed exponents, and $\gamma > 0$ is a constant. The vector u_i is the reflection vector of the velocity v_i from the surface S of the obstacle. It means that if the individual "sees" the obstacle lying on its moving line, and the distance to the obstacle is less than R, it will promptly react for matching its velocity to the reflection vector u_i to avoid the obstacle. Meanwhile, if the distance is larger than R, the reaction is less strong.

The rule for specifying the reflection vector u_i is shown in Fig. 2a for the two-dimensional space and Fig. 2b for the three-dimensional space. Note that in the case where x_C lying on the ray with origin x_i and direction v_i, u_i is simply the vector $-v_i$ starting from the intersection point of the ray with S.

Now, let consider a plane obstacle W, for example, a wall. We specify a plane V that contains the position $x_i(t)$, velocity $v_i(t)$ and orthogonal to the obstacle. Let $l = W \cup V$ be the intersection line of the two plane. Then specifying reflection vector u_i of v_i from the obstacle W becomes to specify the reflection vector of v_i from the line l in the two-dimensional space V (Fig. 2b).

2.4 Foraging Model

We now review a model of the foraging behavior of a group of animals in a noisy environment with obstacles and a food resource. The location of the food resource is fixed in space. The initial position of the individuals is separated from the food resource by obstacles. The separation means that the individuals cannot move to the food resource along any straight lines connecting initial positions to the location of food resource.

We introduce a local rule for foraging as follows. Each individual is sensitive to the gradient of potential formed by the smell of food and has a tendency to move upwards.

Let $f(x)$ be the density function of the food resource defined on the whole domain space Ω. The food resource impacts the movement of individual through its smell. Let $U(x)$ denotes the smell from food resource at position x. It satisfies an elliptic equation (diffusion equation) in Ω under the Neumann boundary condition on $\partial\Omega$:

$$\begin{cases} -c\Delta U + aU = f(x), & x \in \Omega, \\ \dfrac{\partial U}{\partial \mathbf{n}} = 0, & x \in \partial\Omega. \end{cases} \tag{3}$$

Here the notation Δ denotes the Laplace operator $\Delta u = \sum_{k=1}^{d} \frac{\partial^2 u}{\partial x_k^2}$; $c > 0$ is the diffusion constant, $a > 0$ is the declining rate of $U(x)$. The parameters c and a can also be functions of position x and time t. In the Neumann boundary equation, $\mathbf{n}$ denotes the (typically exterior) normal to the boundary $\partial\Omega$. The boundary condition ensures that the domain is perfectly insulated; that is, the food smell cannot pass through the boundary of the domain.

Food smell will guide individuals move toward food's location along its concentration gradient. More precisely, we model the food resource's influence through the external force F_i $(i = 1, 2, \ldots, N)$ as

$$F_i(x_i, v_i) = k\nabla U(x_i), \qquad i = 1, 2 \ldots N, \tag{4}$$

where $k > 0$ is the smell attractive coefficient and ∇ is the gradient notation. Consider C^1 potential function $u : \mathbb{R}^d \to \mathbb{R}$,

$$\nabla u(x) = \left(\frac{\partial u}{\partial x_1}, \frac{\partial u}{\partial x_2}, \ldots, \frac{\partial u}{\partial x_d}\right), \quad x \in \mathbb{R}^d.$$

This force pushes individuals toward the maximum of the potential.

For a technical point, we included a parameter $v_{\max}$ to restrict the maximum speed of individual. If the magnitude of the velocity v_i exceeds $v_{\max}$, we will rescale it to magnitude $v_{\max}$ while preserving its direction. This is reasonable because each species has a tolerance of speed that cannot be exceeded. That is,

$$\hat{v}_i(t) = \begin{cases} v_i(t), & \text{if} \|v_i(t)\| \leqslant v_{\max}, \\ \dfrac{v_i(t)}{\|v_i(t)\|} v_{\max} & \text{otherwise.} \end{cases}$$

3 Numerical Study on Swarming Models

3.1 Swarm Cohesiveness

In this subsection, we review our scientific definition of swarm cohesiveness which characterizes the strength of the group to form and maintain their swarm against noises (Nguyen et al. 2016).

Before going any further, we need some concrete notion of swarming, how can we judge if the group has formed a swarm or not. We introduce such notation using the language of graph theory.

Definition 1 *(ε-graph)* Let $\varepsilon > 0$ be a fixed length. Define $G(t, V(t), E(t))$ be a time-dependent graph. The graph vertice set $V(t)$ at time t is all the positions of particles, $x_i(t)$, $1 \leqslant i \leqslant N$. Two vertices $x_i(t)$ and $x_j(t)$ are connected by an edge in $E(t)$ if $\|x_i(t) - x_j(t)\| \leqslant \varepsilon$. This graph is called the ε-graph of the group at time t.

We also denote by $N_\varepsilon(t)$ the number of connected components of the graph.

The mathematical definition of swarming with distance ε and tolerance speed difference θ is defined as

Definition 2 *(ε, θ-Swarming)* For a given length $\varepsilon > 0$ and a tolerance $\theta > 0$, the group is said to be in ε, θ-*swarming* if there exists a time $T > 0$ such that for all $t \geqslant T$,

$$N_\varepsilon(t) = 1,$$

$$\sigma \mathrm{VS}(t) = \sqrt{\frac{1}{N}\sum_{i=1}^{N} \|v_i(t) - \bar{v}(t)\|^2} \leqslant \theta,$$

where $\bar{v}(t) = \frac{1}{N}\sum\limits_{i=1}^{N} v_i(t)$ is the average of all velocities of individuals at time t.

It is observed that ε, θ-swarming can be formed easily provided that noise magnitudes $\sigma_i(t)$ are not too large.

Figure 1a demonstrates the evolution of forming swarm according to time. The fourth graph depicts the variance of velocity and swarm diameter which is defined as

$$\delta S(t) = \sup_{1 \leq i \leq N} \|x_i(t) - \bar{x}(t)\|, \qquad 0 < t < \infty,$$

where $\bar{x}(t) = \frac{1}{N}\sum_{i=1}^{N} x_i(t)$ is the center of the group at time t.

It is well known that all biological dynamical systems evolve under stochastic forces. It is therefore essential to understand and investigate the influence of noise in the dynamics. In some cases, the noise simply blurs the dynamics without quantitative

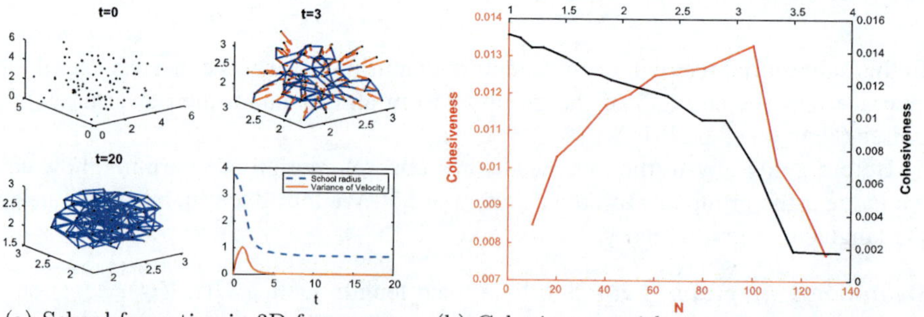

(a) School formation in 3D free space (b) Cohesiveness with respect to parameters

Fig. 1 Simulation results for free space model

effects. However, in a nonlinear dynamical system, the noise drastically changes the corresponding deterministic dynamic behavior of the system.

We are now ready to state the scientific definition for **swarm cohesiveness** that is an internal property characterizing the strength of swarming.

Definition 3 *(Swarm cohesiveness)* Swarm cohesiveness is the ability of a group of animal to form and maintain the ε, θ-swarming structure against the noise imposed on the swarm. In other words, how far the group maintains ε, θ-swarming as the magnitudes of the noises increase.

This definition is given in a quantitative form. When ε and θ are specified, it is possible to quantitatively measure the cohesiveness of a group by means of numerical methods. Let us next give two examples.

Example 1 Consider a group of 50 individuals moving in $\mathbb{R}^2$. The parameters are set as $p = 4, q = 5, \alpha = 4, \beta = 1$. The external force functions are taken as $F_i = -v_i$ for $i = 1, 2, \ldots, N$. Initial positions $x_i(0)$ are randomly located in a suitably small domain with null initial velocities $v_i(0) = \mathbf{0}$ for all $i = 1, 2, \ldots, N$. The magnitude $\sigma_i = \sigma$ is a control parameter of simulation.

We pick out 20 different trajectories of the Wiener process. For each value σ, numerical computations for the solution $x_i(t)$ and $v_i(t)$ are performed in 20 trials corresponding to these trajectories.

Set $\varepsilon = r = 0.5$ and $\theta = 0.5$. It is examined whether or not the states $(x_i(t), v_i(t))$, $i = 1, 2, \ldots, N$ are in ε, θ-swarming by fixing $T = 15$. The algorithm for checking if a group form a swarming for a given parameter set is described in Algorithm 1.

Starting with sufficiently small σ then increase it with increment step 0.001. When the ε, θ-swarming structure is broken down at least for one sample trajectory of the Wiener process, the group considered to lose the ability to swarm. The swarm cohesiveness is the largest value of σ such that the group is still in ε, θ-swarming.

Using this method, we can specify the effect of the model parameters on swarm cohesiveness. We also examine the effect of exponential p and critical radius r on swarm cohesiveness.

For numerical simulation on the effect of p, we tune p as $p = 2, 3, 3.62, 4$, keeping the relation $q = p + 1$ and other parameters as set as above.

In order to calculate swarm cohesiveness according to r, we tune r from 0.5, 0.6 to 0.7 and take $\varepsilon = r, p = 4, q = 5$.

The result is shown in Table 1. The exponent p, as explained in Sect. 2, shows a degree of range how far the attraction is effective. It is therefore very natural that as p decreases, the attraction range extends and enhances the cohesiveness of the group. The cohesiveness is also enhanced as the critical distance increases.

Table 1 Dependence of swarm cohesiveness on model parameters

p	2	3	3.62	4
Cohesiveness	0.063	0.056	0.055	0.051
r	0.5	0.6	0.7	
Cohesiveness	0.051	0.053	0.054	

Algorithm 1 *Swarming Check*

procedure SWARMING CHECK ALGORITHM(σ)
  Input: $d, N, p, r, \alpha, \beta, c, \epsilon, \theta$,
  κ: *numbers of trials,*
  $[T_0, T_1]$: *interval for checking swarming condition*
  Output: swarmingcheck=*true/false*
  swarmingcheck=*true*
  for $k = 1 : \kappa$ ***do***
    if $\exists t \in [T_0, T_1]$ *such that* $N_\epsilon(t) \geqslant 1$ *or* $\sigma \mathrm{VS}(t) \geqslant \theta$ ***then***
      swarmingcheck=false
      Escape from for loop
    end if
  end for
end procedure

Example 2 In this numerical example, we calculate the swarm cohesiveness in three-dimensional space when turning only one parameter N, p, while keep all the others constant as $\alpha = 2$, $\beta = 1$, $c = 0.5$, $T = 2.5$ (i.e., from step 2500), $\varepsilon = 0.5$, $\theta = 0.01$. We use a recursive midpoint-liked algorithm (Algorithm 2) to specify the swarm cohesiveness with 20 arbitrary trails corresponding to different realizations of noises for each parameter set. The two endpoints of the segment to find the swarm cohesiveness are $a = 0$, $b = 0.2$, and error threshold $\delta = 0.0002$. The numerical result is shown in Fig. 1b.

Algorithm 2 *Finding Cohesivness*

procedure FINDING COHESIVNESS ALGORITHM(*startpoint,endpoint,threshold*)
  Input: $d, N, p, r, \alpha, \beta, c, \epsilon, \theta$,
  startingpoint $= 0$,
  endpoint: SwarmingCheck($endpoint$) $= false$
  threshold: allowed error
  Output: Swarm cohesiveness
  $\sigma = (startpoint + endpoint)/2$
  if $(endpoint - startpoint) \leqslant 2 * threshold$ ***then***
    if SwarmingCheck(σ) $== true$ ***then***
      Swarmcohesiveness $= \sigma$
    else
      Swarmcohesiveness $= startpoint$

```
        end if
    else
        if SwarmingCheck(σ) == true then
            startpoint = σ
        else
            endpoint = σ
        end if
        Finding Cohesivness(startpoint,endpoint,threshold)
    end if
end procedure
```

Remark 1 From the above two examples, we see that the swarm cohesiveness increases as critical distance r increases or power p decreases. It is interesting that the cohesiveness increases as the population size increases to some optimal value, and then start to decrease when N excesses that value. This can be explained as follows. It will take more time for a more crowded group of individuals get to the swarming state. Therefore, as we fix the allotted time from which the conditions for swarming must be satisfied, the cohesiveness decreases.

In the following subsection, we show that swarm cohesiveness can be measured quantitatively through obstacle avoidance patterns.

3.2 *Obstacle Avoiding Behavioral Patterns*

We find that there are clear four avoidance patterns and that the emerging pattern changes depending on parameter tuning.

Figure 3a shows the numerical results for p = 2, 3, 3.62, 4 in two-dimensional case. Four different kinds of avoiding patterns are found. We will call them, Rebound (Pattern I), Pullback (Pattern II), Pass and Reunion (Pattern III), and Separation (Pattern IV), respectively. Let us describe these four patterns of swarming.

- **Pattern I (Rebound)**: The individuals keep swarming throughout the obstacle-avoiding process and the swarm rebounds off the obstacle. In order to keep swarm structure, they change their directions after the swarm touch the obstacle.
- **Pattern II (Pullback)**: The individuals are once separated while approaching the obstacle and stay around the surface of the obstacle for a while. They then pullback off the obstacle to reform a swarm structure.
- **Pattern III (Pass and Reunion)**: The individuals pass the obstacle by splitting to move along the obstacle surface. After passing it, they reunite into a single swarm.
- **Pattern IV (Separation)**: It is similar to Pattern III. But, after passing the obstacle, the subgroups have their own directions.

As mentioned above, p is expected to measure a degree of attraction range. Thereby, if the range is long, then the swarm acts as a single living thing. On the contrary, if it is short, then the swarm is easily separated by the obstacle. Tuning the parameter p yields in this way a clear change of patterns in avoiding the obstacle as a swarm. We also choose critical distance r as a tuning parameter. For this case, as r increases, the patterns change gradually in the reverse order.

In the preceding subsection, we have already verified that when other parameters are fixed, the swarm cohesiveness increases as p decreases. When $p = 2$, the swarm has very strong cohesiveness and rebounds off the obstacle. When $p = 3$, the swarm has still strong cohesiveness and can keep swarming but the individuals are spread on the surface. When $p = 3.7$, the swarm cohesiveness becomes weak and the group can no longer keep swarm structure but it is strong enough to reunite the members into a swarm. When $p = 4$, the group cannot keep swarming and is separated into two subgroups after passing the obstacle.

If these interpretations are reasonable, we can use obstacle avoidance patterns in order to measure the swarm cohesiveness easily. More precisely, we can quickly categorize it into four classes. Next, we will investigate more precisely the relations of $p,\ q,\ r$, and the obstacle avoidance patterns.

Example 3 We set the parameters as follows, $d = 2$, $N = 20$, $\alpha = \beta = \gamma = 1$. The exponent p is tuned from 2 to 8 keeping always the relation $q = p + 1$ and $\sigma_i = 0$ for all i. The critical distance is set by $r = 0.5$, and the radius of the obstacle is $\rho = 1.2$. By performing preliminary computations, we first set a stationary state which is in $r,\ 10^{-6}$-swarming whose center is 5 (length distance unit) far from the center of the obstacle. The initial velocities are $\mathbf{v}_i(0) = (1.75,\ 0)$ for all i. The parameters for obstacle avoidance are set as $P = p$, $Q = q$, $R = r$. The swarm is oriented toward the obstacle, and after a while, the individuals strike on it.

In Nguyen et al. (2014), we show that as the critical distance increases, the school diameter also increases. However, the school diameter will also affect the school pattern when avoiding obstacle. Therefore, in our numerical simulation to demonstrate the effect of critical radius r to the swarm cohesiveness, we change the radius of obstacle according to the change of critical distance. More precisely, after getting the equilibrium position of fish in a swarm for each simulation that is $r,\ 10^{-6}$-swarming, we take obstacle radius equal to 2/3 of swarm diameter. The distance between the centers of the swarm and the obstacle is 8. In the initial time, all individuals have the same velocity with length 4 and the same direction as the vector which connects the center of school and center of the obstacle. Common parameters are taken as in the simulation for the effect of power p. Besides, $p = 3$.

The numerical result is given in Table 2.

Relationship between behavior patterns, swarm cohesiveness and model parameters

Emerging patterns (patterns I, II, III, IV) can be achieved just by tuning one parameter while keeping all others constant. Moreover, the relationships between model parameters, swarm cohesiveness, and behavioral patterns are shown in the diagram below.

Table 2 Different patterns can be achieved just by tuning one parameter while keeping all others constant

p	[1.001,2.100]	[2.101,3.371]	[3.372,3.497]	[3.498,8]]
Pattern	I	II	III	IV
r	[0.2, 0.3]	[0.4, 0.5]	[0.6, 2.0]	[2.1, 2.8]
Pattern	IV	III	II	I

- p, q increase $\Rightarrow$ cohesiveness decreases $\Rightarrow$ behavioral pattern index increases,
- r increases $\Rightarrow$ cohesiveness increases $\Rightarrow$ behavioral pattern index decreases.

3.3 *Foraging Advantage*

This subsection presents a numerical result that is consistent with many observations on foraging process of swarm in the real world.

Consider the moving of animals on the ground. In our simulations, we restrict the space for individuals moving in to be the rectangle domain $D = [0, 7] \times [0, 4]$ whose boundary are wall obstacles. Inside this domain, we put three obstacles

$$\begin{aligned} \mathrm{Ob}_1 &= [2,\ 2.5] \times [1.75,\ 4] \\ \mathrm{Ob}_2 &= [4.5,\ 5] \times [0,\ 2.5] \\ \mathrm{Ob}_3 &= \{(x, y) \in \mathbb{R}^2 : x = 2.25 + 0.25\cos\theta,\ y = 1.75 + 0.25\sin\theta, \theta \in [-\pi, 0]\}. \end{aligned}$$

The third obstacle is the below half of the circle centered at [2.25, 1.75] with radius 0.25. It is put right after the first obstacle to make one block. Therefore, the area where the individuals can move in is the domain

$$\Omega = D\backslash(\mathrm{Ob}_1 \cup \mathrm{Ob}_2 \cup \mathrm{Ob}_3).$$

By the Neumann condition, smell also cannot pass the boundary $\partial\Omega$.

The initial position of the individuals is taking randomly on a small rectangle on the upper left corner of the domain Ω. Figure 2c (left) shows the environmental configuration of the simulations.

We set the moving environment, food resource, and initial positions of individual as in Sect. 2.4. There is a circle food resource centered at $C = (5.5, 0.1)$ with radius 0.04. Precisely, we specify the food resource function as

$$f(x) = \begin{cases} 50, & \text{if} x \in \{y \in \mathbb{R}^2 : \|x - C\| \leqslant 0.04\} \\ 0 & \text{else.} \end{cases}$$

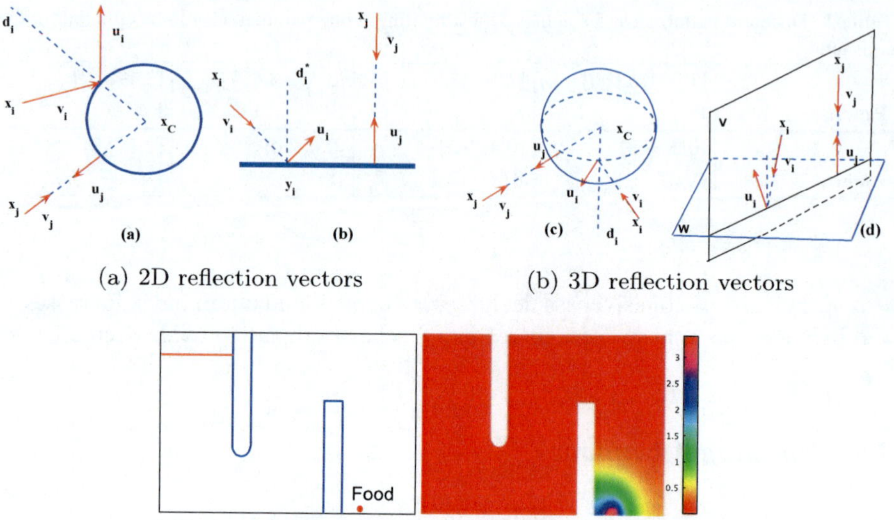

(a) 2D reflection vectors

(b) 3D reflection vectors

(c) (left) Configuration of the foraging model simulation. The position of the individuals at the initial time is taken randomly in the small upper-left rectangle. (right) Food smell potential function

Fig. 2 Reflection rules in 2D, 3D, and environment setting for foraging model

The coefficients for specifying the smell emitting from food in (3) are $c = 0.1$, $a = 0.2$. The food coefficient in (4) is taken $k = 20$.

We say that the group successes in finding the food resource if all the individuals pass the second obstacle within the allotted time $T = 300$, that is before the 300,000th iterative is reached, as we take the step size for discrete scheme $\Delta t = 0.001$. Figure 3b illustrates an evolution of finding food resource.

Other parameters are set as $\alpha = 2$, $\beta = 1$, $\gamma = 5$, $p = P = 3$, $q = Q = 5$, $r = 0.1$, $R = 0.3$, $\sigma = 0.001$, $v_{\max} = 0.8$.

The numerical result is shown in Fig. 3c. We can see that animals can enjoy a swarming advantage when finding food resources. A single individual hardly finds the food resource. The probability of successful foraging increases as the population increases to some optimal value and decreases as population size exceeds that value. This result agrees with many experimental observations (Berdahl et al. 2013; Gotmark et al. 1986; Steele et al. 1991). The existence of such an optimal value may be explained by swarm cohesiveness (see Remark 1). Furthermore, the averaging time for successful foraging at that optimal value is also the smallest.

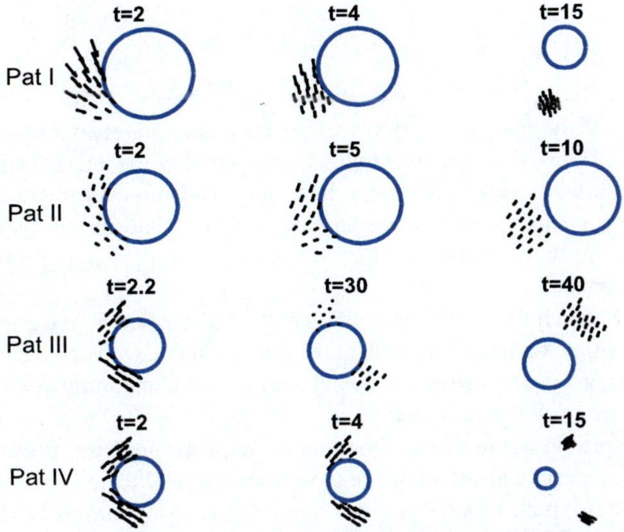

(a) Pattern changes as parameter p is tuning

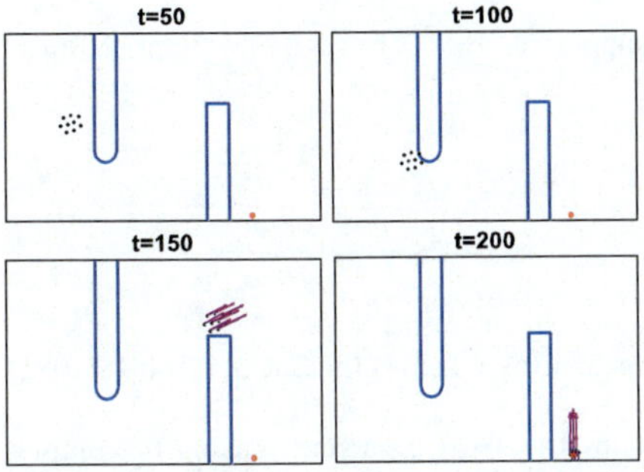

(b) Evolution of finding food resource

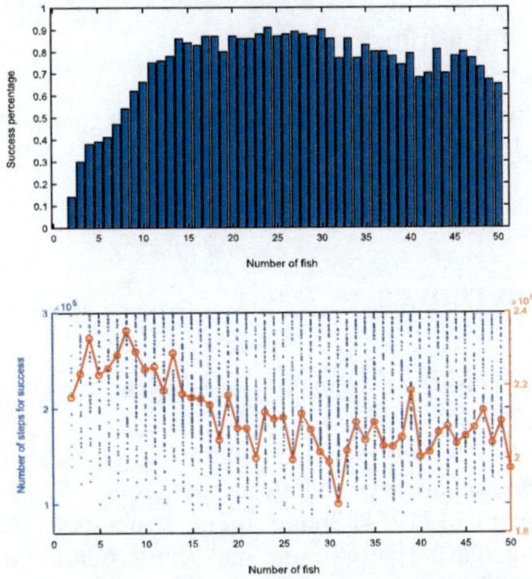

(c) (above) Experimental probability of foraging success (based on 100 trials at allotted time $T = 300$) with respect to population size, (below) simulation times (point) for succuessful trials and average of successful times over trials

Fig. 3 Simulation results for avoiding obstacle model and foraging model

4 Conclusion

We proposed some SDEs describing the swarming of animals in different environments. We introduced the so-called ϵ, θ-swarming for quantitatively defining swarming and a scientific notion of cohesiveness which characterizes the internal strength of the swarm. We observed four obstacle-avoiding patterns. Moreover, there are close relationships between these patterns, swarm cohesiveness, and model parameters. Also by numerical study, we confirmed that animals can enjoy foraging advantages while forming swarms.

Our results may have important implications for the development of technologies, such as swarm robotics, with applications in the detection of explosives, landmines, or people in search-and-rescue operations. They may be used for computer animation, visualizing information or optimization.

One interesting problem is to extend the model to describe predator–prey systems, in which the predators apply an attack strategy for foraging while the prey chooses an escape strategy for survival. Empirical observations show that animals benefit from swarming behaviors when foraging and escaping enemies. It is possible to construct effective attack strategies and escape strategies using mathematical models. This is left for a future work.

Acknowledgements This work was partially supported by JST CREST Grant Number JPMJCR14D6, Japan.

References

Aoki I (1982) A simulation study on the schooling mechanism in fish. B Jpn Soc Sci Fish 48:1081–1088

Berdahl A, Torney CJ, Ioannou CC, Faria JJ, Couzin ID (2013) Emergent sensing of complex environments by mobile animal groups. Science 339:574–576

Breder CM (1951) Studies on the structure of the fish school. Bull Smer Mus Nat Hist 98:1–28

Breder CM (1959) Studies on social grouping in fishes. Bull Amer Mus Nat Hist 117:393–482

Camazine S, Deneubourg JL, Franks NR, Sneyd J, Theraulaz G, Bonabeau E (2001) Self-organization in biological system. In: Princeton University Press

Carron AV (1980) Social psychology of sport, vol 31. Mouvement Publications, New York

Cullen JM, Shaw E, Baldwin HA (1965) Methods for measuring the three-dimensional structure of fish schools. Anim Behav 13:534–543

D'Orsogna MR, Chuang Y, Bertozzi A, Chayes L (2006) Self-propelled particles with soft-core interactions: patterns, stability and collapse. Phys Rev Lett 96:104302

Festinger L, Schachter S, Back K (1950) Social pressures in informal groups. Harper and Row, New York

Gotmark F, Winkler DW, Anderson M (1986) Flock feeding on fish schools increases individual success in gulls. Nature 319:589–591

Gross N, Martin WE (1952) On group cohesiveness. Am J Sociol 57:546–564

Huth A, Wissel C (1992) The simulation of the movement of fish school. J Theor Biol 156:365–385

Kloeden PE, Platen E (2005) Numerical solution of stochastic differential equations. Springer

Keenleyside M (1995) Some aspects of the schooling behaviors of fish. Behaviour 8:183–248

Nguyen LTH, Ta TV, Yagi A (2014) A quantitative investigations for ODE model describing fish schooling. Sci Math Jpn 77:403–413

Nguyen LTH (2014) Numerical study on some stochastic models in biology. In: Doctor's dissertation, Osaka University, Japan

Nguyen LTH, Ta TV, Yagi A (2016) Obstacle avoiding patterns and cohesiveness of fish school. J Theor Biol 406:116–123

Moreno JL, Jennings HH (1937) Statistics of social configurations. Sociometry 1:342–374

Oboshi T, Kato S, Mutoh A, Itoh H (2002) Collective or scattering: evolving schooling behaviors to escape from predator. In Artificial life. MIT Press, Cambridge, VIII, 386–389

Olfati-saber R (2006) Flocking for multi-agent dynamic systems: algorithms and theory. IEEE Trans Automat Control 51:401–420

Parrish JK, Viscido SV (2005) Traffic rules of fish schools; a review of agent-based approaches. In: Cambridge University Press, Cambridge

Reynolds CW (1987) Flocks, herds, and schools: a distributed behavioral model. Comput Graph 21:25–34

Steele C, Scarfe A, Owens D (1991) Effects of group size on the responsiveness of zebrafish, Brachydanio rerio (Hamilton Buchanan), to alanine, a chemical attractant. J Fish Biol 38:553–564

Ta TV, Nguyen LTH (2008) A stochastic differential equation model for the foraging behavior of fish school. Phys Biol 15:036007

Uchitane T, Ta TV , Yagi A (2012) An ordinary differential equation model for fish schooling. Sci Math Jpn 75:339–350, e-2012, 415–426

Vicsek T, Czirók A, Ben-Jacob E, Cohen I, Shochet O (1995) Novel type of phase transition in a system of self-driven particles. Phys Rev Lett 75:1226–1229